Practical Biochemistry

A Comprehensive Guide to Undergraduate Students

Shailaza Shrestha
M.Sc. Medical Biochemistry
Department of Biochemistry
Hind Institute of Medical Sciences
Ataria, Sitapur

PEEPEE
PUBLISHERS AND DISTRIBUTORS (P) LTD.

PRACTICAL BIOCHEMISTRY
A Comprehensive Guide to Undergraduate Students

Published by
Pawaninder P. Vij and Anupam Vij

Peepee Publishers and Distributors (P) Ltd.
Head Office: 160, Shakti Vihar, Pitam Pura
Delhi-110 034 (India)

Correspondence Office:

7/31, First Floor, Ansari Road, Daryaganj, New Delhi-110002 (India)
Ph: 65195868, 23246245, 9811156083
e-mail: peepee160@yahoo.co.in
e-mail: peepee160@rediffmail.com
e-mail: peepee160@gmail.com
www.peepeepub.com

First Edition: 2016

ISBN: 978-81-8445-203-7

I dedicate this book to my Parents

(Mr. Shiva Narayan Shrestha and
Mrs. Buddha Laxmi Shrestha)

and

Students

Preface

This practical book is written with the main focus towards undergraduate students in order to provide a comprehensive knowledge to them. The topics covered include qualitative experiments of carbohydrates, proteins, lipids and urine, quantitative estimations of various analytes present in blood, demonstration practical, case studies and spotters with answers. Viva questions, reagent preparation methods and reference range of several biochemical parameters are also included. The content of the book is made as simple as possible so that this will help the students on understanding biochemistry in much easier way such that it can be fruitful to them not only in their studies but also in their medical career in future.

Shailaza Shrestha

Acknowledgements

I am very glad that I am able to write this book. But the effort made was not by myself alone. Lots of helping hands were there to support me. I take this opportunity to thank my students who encouraged me to write the book. I am very grateful to Mr. Rahul Rai for his immense support during the preparation of the manuscript. I would also like to express my gratitude to Dr. Md. Tabrej Alam, Mr Mukesh Kumar Singh and Ms. Mahak Jain for providing me the new ideas to improve the manuscript. I am highly thankful to my Head of the Department, Dr. D Suryanarayana, seniors Dr. Saleha Shaheen, Dr. Sadhana Ajay and my fellow colleagues for always being with me and supporting me.

I gratefully acknowledge Dr. A.K. Srivastava (Vice Principal and Head of the Department of Anatomy) for guiding and helping me to publish this book.

Shailaza Shrestha

Contents

1

LABORATORY HAZARDS WITH SAFETY RULES

Biochemistry laboratory deals with various tests and investigations, so one must be alert and cautious while working in the laboratory. Students must be aware of all the potential hazards that can occur due to careless handling of chemicals, specimens or glassware.

Laboratory Hazards may be due to:

- Broken glassware.
- Poisonous fumes.
- Toxic and inflammable chemicals.
- Electric and mechanical causes.

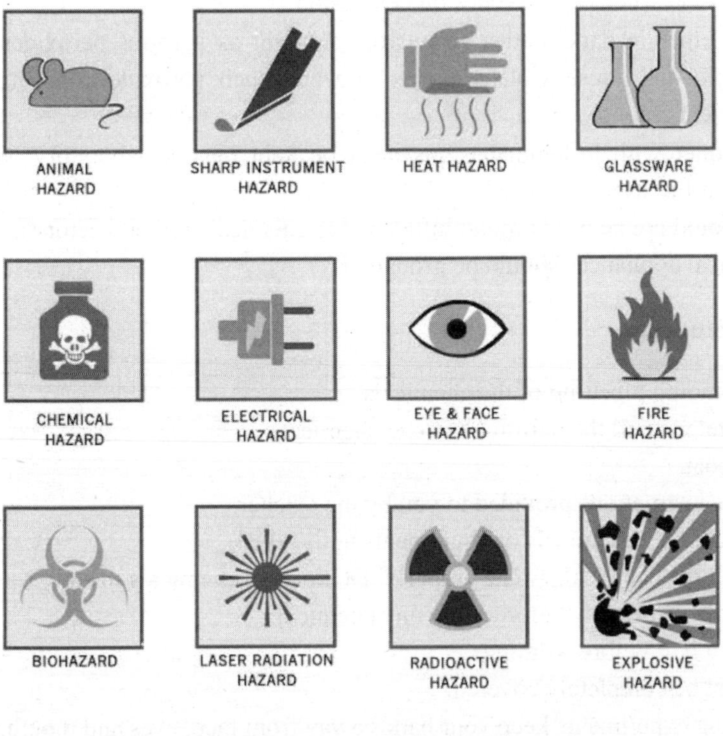

Fig. 1.1: Various types of laboratory hazards

PRECAUTIONS TO AVOID HAZARDS

For Reagents and Chemicals

1. Laboratory coat is mandatory.
2. Label all the chemicals and reagents.
3. Hold reagents bottles with both hands.
4. Corrosive chemicals must be handled with great care. One should be careful while opening bottles containing such chemicals or while pouring them. Pour them slowly to water with continuous cooling and stirring, e.g., Strong acid and Bases.
5. Lock up the toxic chemicals in cupboard.
6. Carcinogenic chemicals like benzidine, O-toulidine must be labeled properly and their contact to skin should be avoided.
7. Wear rubber or plastic gloves while handling toxic and carcinogenic chemicals.In case of accidental contact wash with water several times.
8. Handle explosive substances with proper care e.g.:
   - Work with perchloric acid in fume cupboard.
   - Store picric acid in container with water and close tightly with cork or rubber stopper as picric acid when dry explodes on percussion.
   - Keep ether in dark bottles away from sunlight as it forms peroxides when exposed to sunlight and cause violent explosion when such peroxide concentration increases to sufficient amount.
   - Container with inflammable gas must be kept outside laboratory when not in use to prevent fire hazards.
9. Burner should not be used around inflammable chemicals (ether, acetone).
10. All electrical appliances should be grounded.

Safety Rules for Students

1. Never do mouth pipetting of the chemicals.
2. Never taste or smell the harmful or toxic chemicals.
3. Wear labcoat.
4. Follow the instructions provided to you by the teacher.
5. Do not touch any chemicals or equipments until asked.
6. Dispose all the chemicals waste properly. Do not mix chemicals in sink.
7. Read the label correctly before using any chemical.
8. Tie up long hair before working.
9. Shoes must be completely covered.
10. While doing experiments keep your hands away from face, eyes and mouth.

11. Never use broken, cracked or chipped glassware.
12. Keep your working are clean.
13. Wash all the glasswares after using them and keep properly on racks.
14. Keep the reagents in proper order in shelves.
15. Wash hands with soap and water after working.
16. Eating and drinking in the laboratory must be avoided.

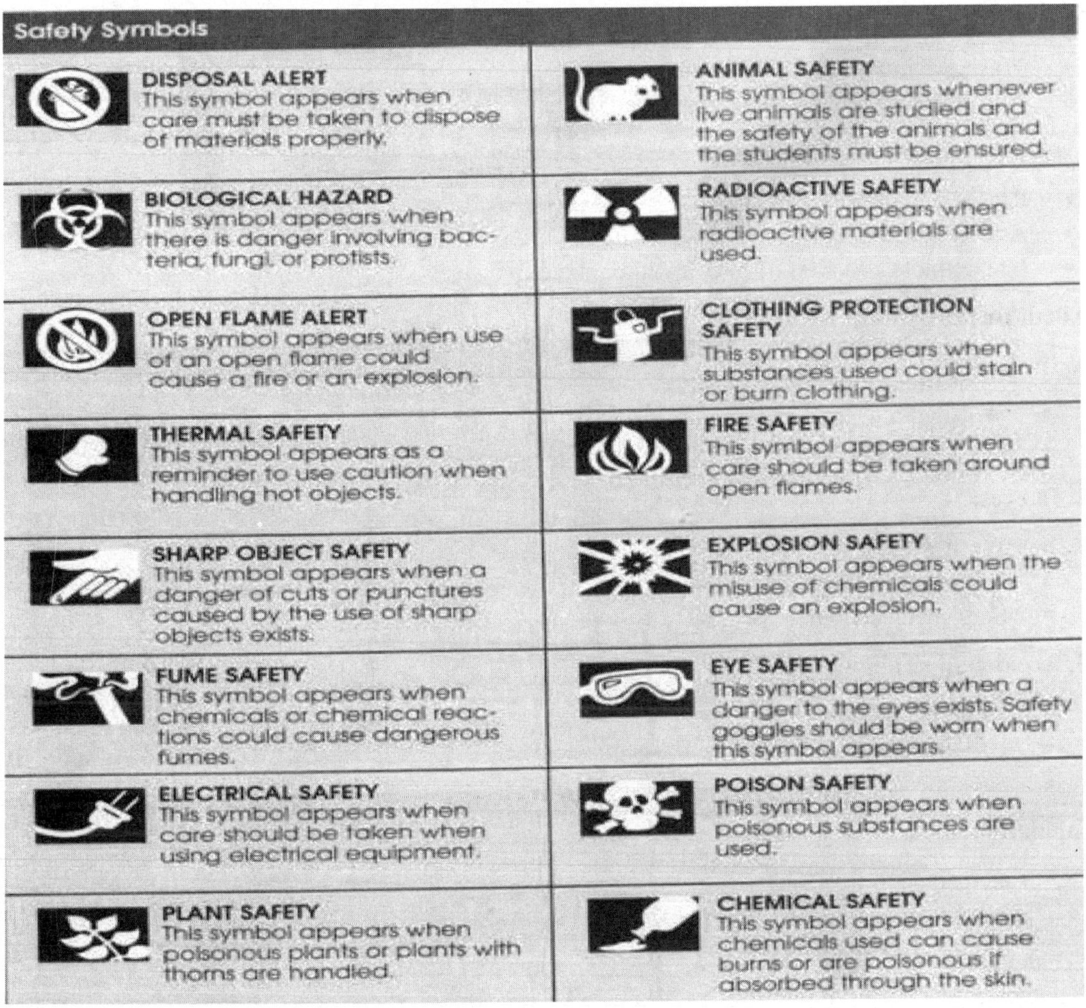

Fig. 1.2: Safety symbol

FIRST AID FOR ACCIDENTS AND INJURIES

A. Acid Burns (HNO_3, H_2SO_4, HCL):

1. In skin:

- Wash the affected area first with water.
- Wash with 5% Na_2CO_3 or NH_4OH.

2. In eyes:

- Wash eye immediately with water.
- Add 2% $NaHCO_3$ solutions. Seek medical help.

3. Accidental swallowing of acid:

- Rinse the mouth.
- Ask the person to drink 5% soap solutions.
- Use antidote (8% MgOH or egg white mixed in water).

B. Alkali Burns (NaOH, KOH, NH_4OH):

1. In skin:

- Wash with water.
- Use 5% acetic acid.

2. In eyes:

- Wash with large amount of water.
- Use saturated solution of boric acid.
- Seek medical help.

3. Accidental swallowing of alkali:

- Rinse the mouth.
- Use antidote (lemon juice or 5% acetic acid).
- Seek medical help.

C. Inhalation of corrosive gases:

- Take person to fresh air.
- Seek medical help.

D. Accidents due to broken glassware:

- Wash the wound.
- Remove any glass pieces.
- Apply ointment and seal with adhesive tape.
- Seek medical help if cut is deep.

2

SAMPLE COLLECTION AND PROCESSING

For diagnosis of various diseases biological samples are collected. These samples include blood, CSF, saliva, urine, pleural fluid, sweat etc. Analysis of these samples can be done by various semi-quantitative and quantitative techniques. Most of the diagnosis tests are carried out in blood sample. Tests are to be performed as soon as possible depending upon the stability of analyte *in vitro* (not more than 1-2 hours after collection). If there is any delay then samples should be stored under suitable conditions. The samples must be refrigerated at 4°C or -20°C depending upon the delay.

Collection of Blood

It can be done by three techniques:

1. Venipuncture (most common).
2. Arterial puncture.
3. Capillary puncture.
   - Puncture area is sterilized by spirit, 70% alcohol or ether.
   - Venipuncture is mainly done on ante cubital vein.
   - Arterial puncture is commonly done in radial or femoral artery.
   - Capillary puncture is done for smear preparation or dry chemistry. It is done at tip of middle finger, ball of great toe, edge of knee newborn.
   - Hemolyzed sample is unsuitable for biochemical estimation, so care should be taken to avoid hemolysis.

Anticoagulant

- Blood specimens used for biochemical investigation may be whole blood, serum or plasma. It depends upon the biochemical parameters investigated and method used.

  Blood-blood cells = plasma

  Plasma-fibrinogen = serum
- For collection of serum, whole blood is allowed to clot for 20-30 minutes at room temperature. It is then centrifuged (3000-5000 rpm for 5 minutes) and serum is collected for analysis.
- For collecting plasma, anticoagulant is used.

- Commonly used anticoagulants area: Potassium oxalate, heparin, ethylene diamine tetra-acetate (EDTA), sodium citrate etc.
- Use of anticoagulant depends on the nature of analyte to be investigated, e.g., Oxalate should be used in blood sample collected for Ca^{2+} investigations. For the sample used for blood glucose estimation, in addition to oxalate, sodium fluoride is also used which is an inhibitor of endase enzyme of glycolysis. Enzyme inhibition prevents glucose utilization by RBC during processing and hence aid in proper estimation blood glucose.

Collection of Urine Samples

Early morning urine sample is required for qualitative investigations whereas for quantitative ones 24 hours urine sample is required. Urine should be collected in clean and dry containers.

Storage of Urine Samples

Precipitation of phosphates, crystalization of uric acid, bacterial decomposition etc may occur in case of improper storage. So preservatives are to be used for proper storage of urine samples. Urine should be collected under sterile conditions. Bacterial contamination causes decomposition of urine to ammonia increasing the urinary pH. Urine also may become alkaline due to precipitation of uric acid and ureates. Thus different preservatives can be used based upon the nature of analysis. Most commonly used preservatives are concentrated HCl, chloroform (to stop bacterial growth), thymol, toluene, liquid petroleum, formalin etc.

Other Samples

1. **CSF:** Test should be done as soon as possible after collection, if any delay; it should be stored at–20°C. Commonly done investigations in CSF are chlorine, sugar protein.
2. **Sweat:** Used for estimation of chloride.

3

GLASSWARE AND INSTRUMENTS

Glasswares used in biochemistry laboratory are mostly made up of borosilicate glass which is resistant to acid, bases corrosion, radiation etc. They used for various purposes like to measure and prepare samples or reagents, to transfer or store the solution etc.

TYPES OF GLASSWARE USED

1. Beaker:

- Used for preparation of samples or reagents.
- Mostly available in ranges like 50 ml to several liters.

2. Measuring cylinder:

- Used to measure specific volumes.
- Ranges from 10 ml-2 liters.

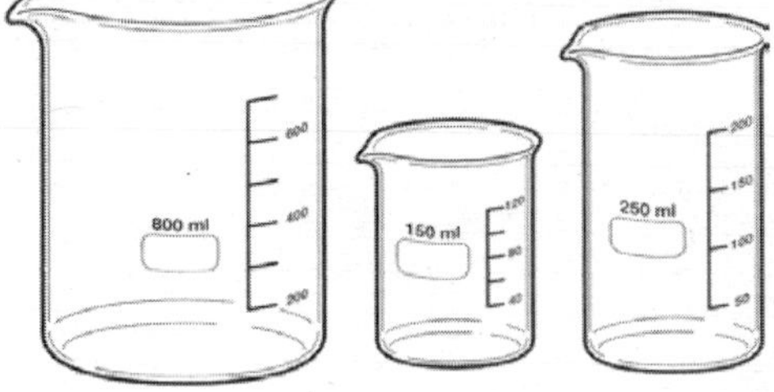

Fig. 3.1: Beaker

Fig. 3.2: Measuring cylinder

3. Volumetric flasks:

- Contain flat bottom with long narrow neck and a stopper.
- Ranges from 10 ml-5 liters.
- 100 ml capacity most commonly used.

4. Conical flasks:

- Used for titration and boiling solutions.
- Evaporation is minimum due to conical shape.
- Ranges from 25-500 ml.

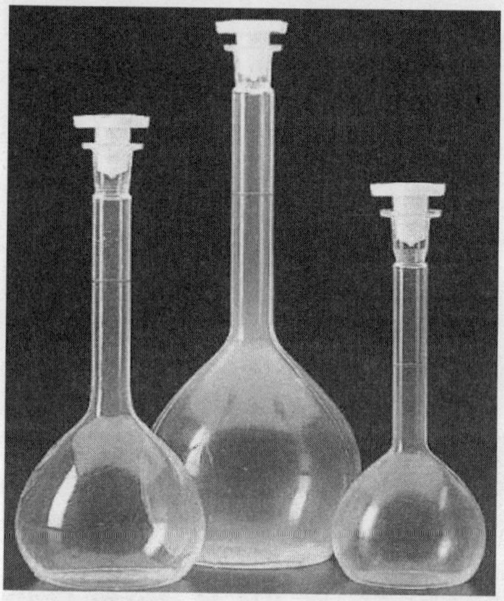

Fig. 3.3: Volumetric flasks

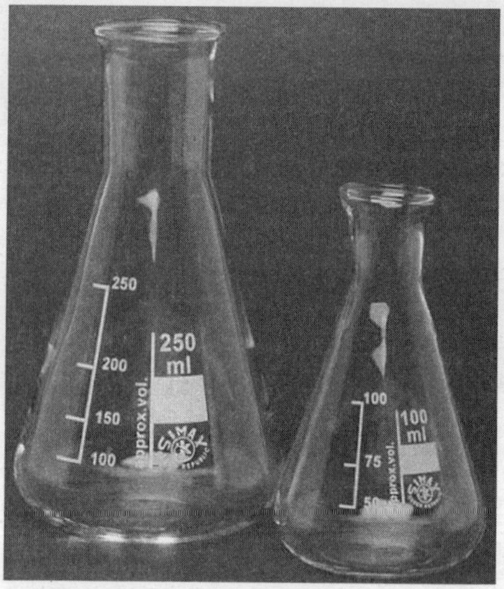

Fig. 3.4: Conical flasks

5. Round bottomed flask:

- Used for distillation.

Fig. 3.5: Volumetric flasks

6. Burettes:

- Long graduated tubes with cork at lower end.
- Capacity ranges from 10 ml-100 ml.
- Used to transfer known volume of liquid.

7. Funnels:

- May be made up of glass or plastic.
- Used for filtration using filter paper.
- Also used to transfer solution to the container with narrow opening (funnel have narrow stem).
- Funnel with wide stem are used to transfer solids to a container.

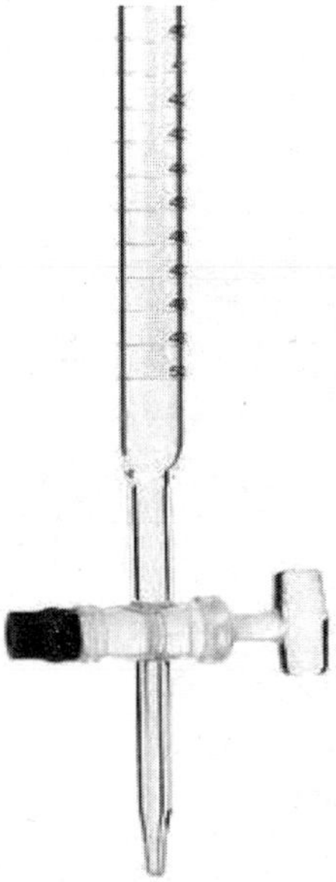

Fig. 3.6: Burette

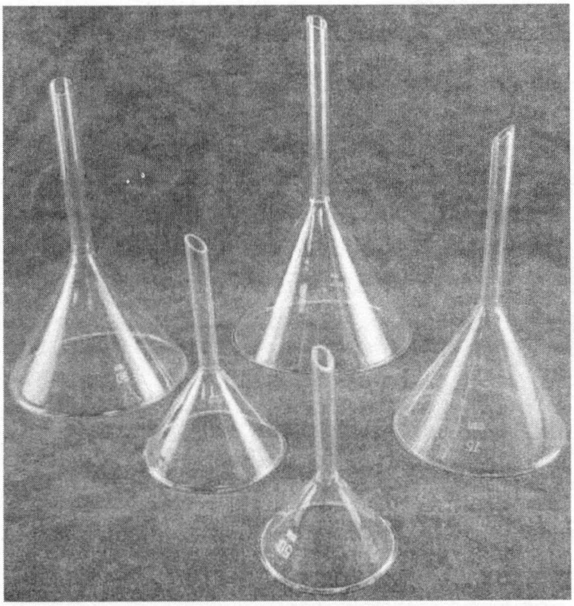

Fig. 3.7: Funnel

8. Pipettes:

i. *Glass pipettes:*

- Graduated long narrow tubes with narrow upper end for pumping the liquid using rubber bulb and gradually tapering, narrow lower end to deliver the liquid.
- Used to transfer accurate volume.
- Glass pipette may be two types:
  - **A. Blow out type:** It contains graduation till the tip of the lower end and the liquid should be blowed out till the last drop.

B. Non blow out type: It does not contain graduation till the tip and the content should be delivered till the last marking (need not to be blow out). The capacity is 1 ml-10 ml.

ii. *Autopipettes:*

- Most commonly used in clinical laboratory.
- Piston operated devices to transfer liquid.
- It may be:

A. Fixed volume type:

- Transfer fixed volume e.g.10 μl, 20 μl, 100 μl, 200 μl, 1000 μl etc.

B. Adjustable type:

- Volume can be adjusted depending upon the need.
- Available in the ranges like 5-20 μl, 20-200 μl, 10-100 μl, 100-1000 μl etc.

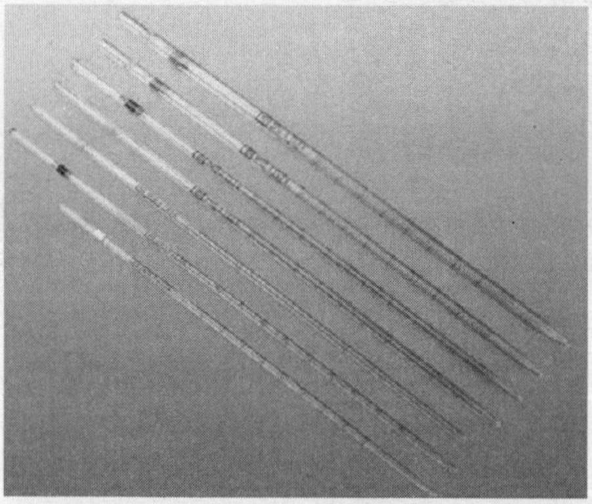

Fig. 3.8: Pipettes

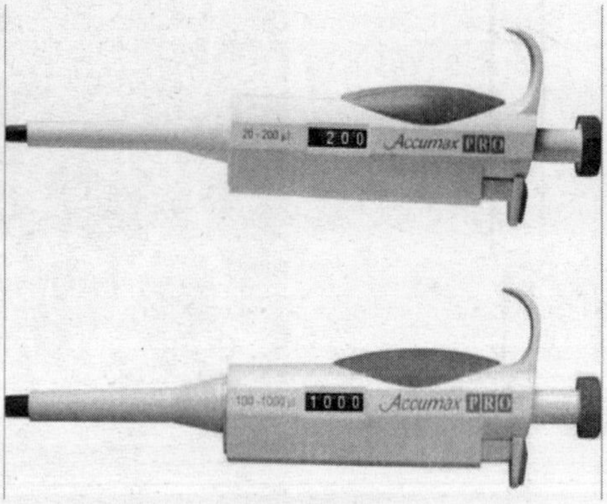

Fig. 3.9: Autopipettes

9. Test tubes:

- Used to perform tests.
- Size available are:
  - ➢ 10 × 15 mm (outer diameter and length)
  - ➢ 15 × 125 mm
  - ➢ 18 × 150 mm

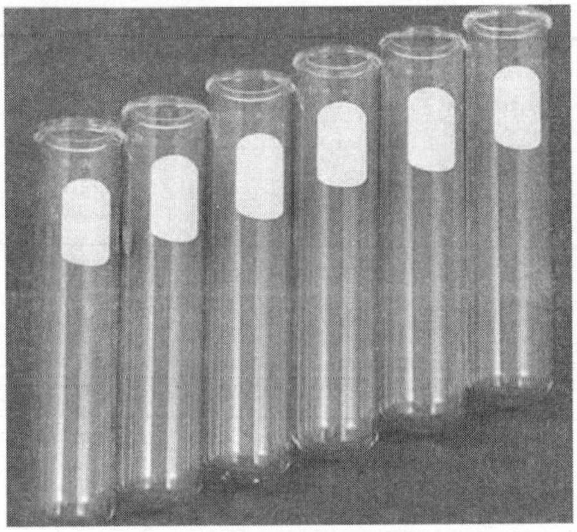

Fig. 3.10:

10. Reagents bottles:

- Cylindrical bottles with narrow neck and stopper.
- Plain bottles are used to store reagents.
- Amber colored bottles are used for light sensitive reagents.
- Capacity–25-200 ml.

11. Desiccators:

- Contain desiccant at the bottom.
- Used to keep solid and liquid substances dry (by removing water).

Fig. 3.11: Reagent bottle

Fig. 3.12: Desiccator

Instruments

Commonly used instruments are:

1. Balances

- Used to measure small quantities accurately (may be in mg or even less).
- They may be:

A. Single pan-chemical balance:

- Most commonly used.
- Highly accurate and sensitive.
- Has taring function which enables the devices to consider weight of the container itself while weighting.

B. Double pan-chemical balance:

- Also called mechanical balance.
- Weighting ranging from 100 mg–several grams can be taken.
- Weight of the container should be taken in account.

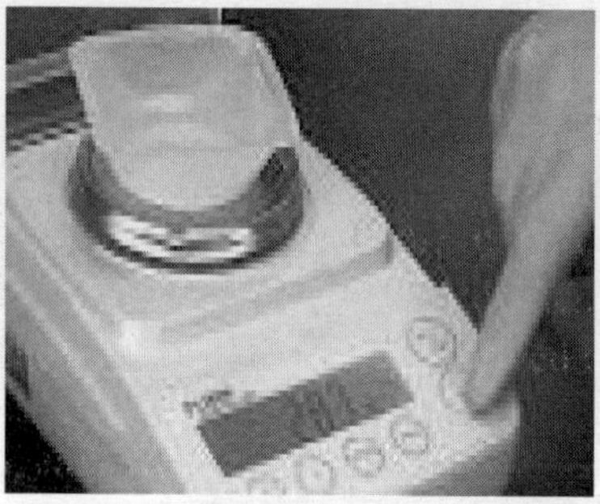

Fig. 3.13

Fig. 3.14

2. Incubator

- Used to provide constant temperature for various biological reactions.
- Contains thermostat to maintain temperature.
- Inner part is made up of stainless steel.
- Incubators available in different sizes.

3. Hot air oven

- Used in laboratory for drying glasswares.
- They have very high temperature with thermostat for temperature regulation.
- Like incubators, hot air oven is also made up of stainless steel body and come in different sizes.

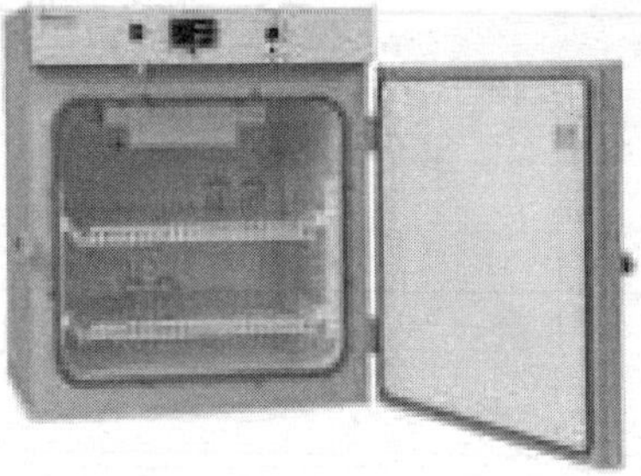

Fig. 3.15: Incubator

Fig. 3.16: Hot air oven

4. pH meter

- Device used to measure the pH of the fluids.
- Contains power pack and two electrodes (glass electrode and calomel electrode).

5. Centrifuge

- Used to separate solute from the solution by rotating the solution at a very high speed.
- Centrifugal force is generated on rotation at very high speed. It depends on mass of the solute particle, radius of rotational path from central axis and speed (rotation per minute).

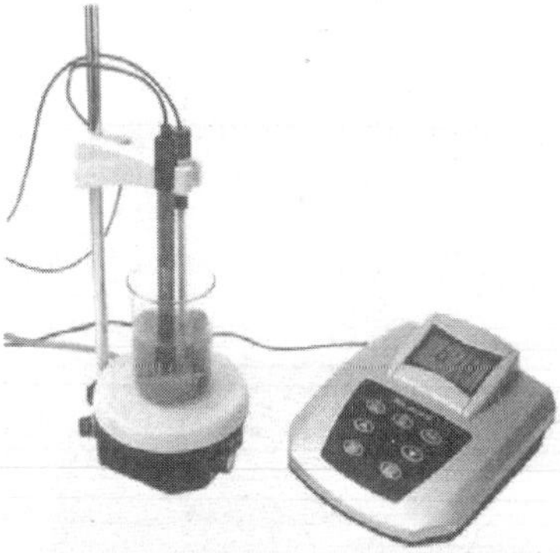

Fig. 3.17: pH meter

Fig. 3.18: Centrifuge

6. Waterbath

- Made up of stainless steel with a heating device and thermaostat for regulating temperature.
- Used for heating purposes (incubation) or to provide boiling temperature for various biochemical reactions.

Fig. 3.19: Waterbath

7. Colorimeter

- Used in quantitative estimation of biological parameters like glucose, protein etc.
- The analyte is first converted to the colored compound by the chemical reaction, then absorbance is taken using colorimeter after selecting suitable filter and wavelength.
- Filter selection depends on the color of the solution (complementary filter is used).
- From the absorbance concentration of unknown analyte can be calculated.
- Works on colored solutions only.
- Follows Beer Lambert law (Light absorbed by the colored solution or absorbance is directly proportional to the concentration of the solute in the sample).

8. Spectrophotometer

- Like colorimeter, it is also used in quantitative estimation of biological parameters
- Both colored and colorless solutions can be used
- Follows Beer Lambert law

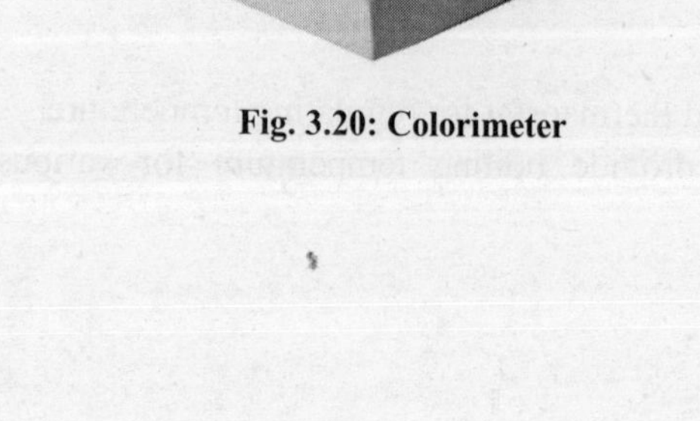

Fig. 3.20: Colorimeter

Fig. 3.21: Spectrophotometer

4

CARBOHYDRATES

Carbohydrates are known as polyhydroxy aldehydes or ketones. They may be classified as:

1. Monosaccharides
2. Oligosaccharides
3. Polysaccharides
   1. **Monosaccharides:** They are also called simple sugar containing a single unit *e.g.* glucose, fructose, mannose, galactose.
   2. **Oligosaccharides:** They are made up of 2-10 monosaccharides *e.g.* lactose, raffinose.
   3. **Polysaccharides:** They are made up of more than 10 monosaccharides unite *e.g.* starch, glycogen.

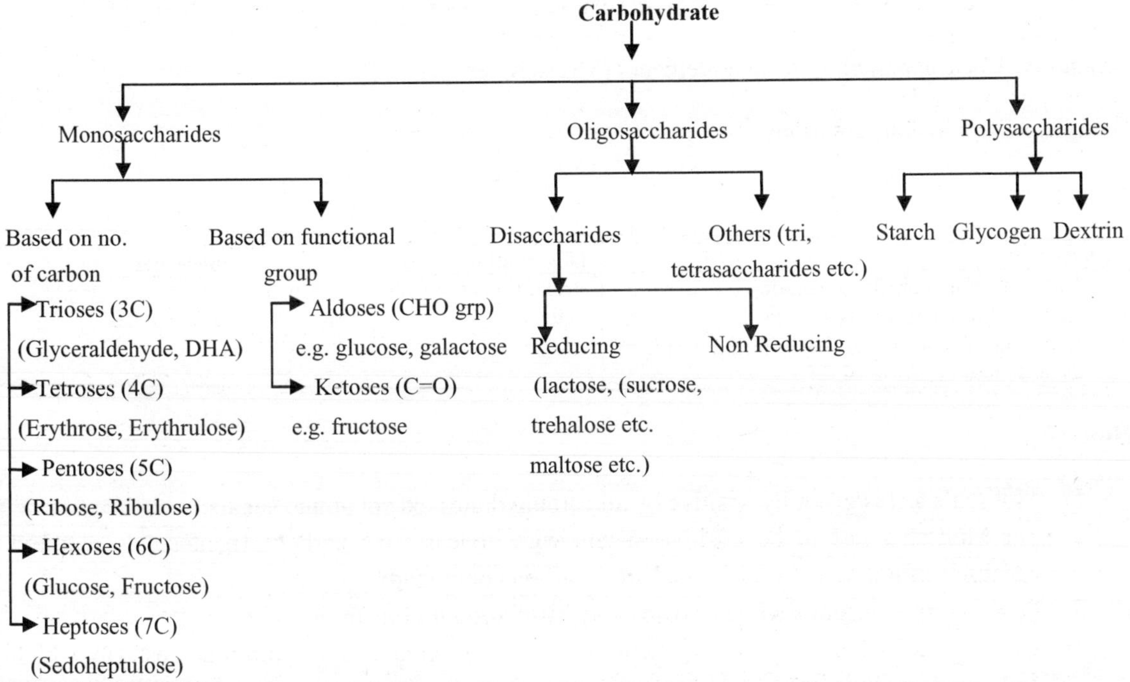

DHA: Dihydroxy acetone

GENERAL REACTIONS FOR CARBOHYDRATES

Molisch's Test

It is the general test for carbohydrates.

Principle

Carbohydrates on treatment with conc H_2SO_4 undergo dehydration (loss of 3 H_2O mol) to produce furfural derivatives. These derivatives then condense with α-naphthol to produce violet colored complex. Diasaccharides and polysaccharides are hydrolysed first to their monosaccharides units and then dehydrated to form furfural.

Reactions

$$\text{Polysaccharide} \xrightarrow{\text{Conc. } H_2SO_4} \text{Monosaccharide.}$$

$$\text{Disaccharide} \xrightarrow{\text{Conc. } H_2SO_4} \text{Monosaccharide.}$$

$$\text{Monosaccharide} \xrightarrow{\text{Conc. } H_2SO_4} \text{Furfral derivatives} \xrightarrow{\alpha\text{-naphthol}} \text{Violet colored complex.}$$

Reagents: 1% α-naphthol in methanol/ethanol (Molisch's reagents)

Samples: Carbohydrate solutions.

Procedure

Experiments	Observations	Inferences
• Take 2 ml of carbohydrates solution • Add 1-2 drops of α-naphthol • Add 2 ml of conc H_2SO_4 slowly from the side of the test tubes	Violet ring at the interface of two liquids	The given sample is carbohydrates

Note:

1. Molisch's test is given by positive by all carbohydrates except amino sugars.
2. For Molisch's test to be positive sugars with at least five carbons (pentose) is required because furfural derivatives formed are 5 carbon compounds.
3. Pentoses give furfurals while hexoses give hydroxymethyl furfural.
4. Besides Molisch's test, other general test for carbohydrates is Anthrone test (in which anthrone is used in place of α-naphthol). Blue color develops.

Precautions:

1. While pouring the acid, first tilt the test tube and add acid slowly from the side of the test tubes.
2. Green color develops if access α-naphthol is added.
3. Black/Brown ring occurs due to charring of sugars by acid.

Iodine Test

This test is given positive by polysaccharides.

Principle

Polysaccharides are large molecules and can adsorb small molecules of iodine on their surface to form polysaccharides iodine adsorption complex.

- Starch gives blue color.
- Glycogen gives reddish brown color.
- Dextrin gives violet color.

The complex formed is dissociated on heating thereby giving colorless solution whereas on cooling color reappears due to reformation of the complex.

Reagents: 0.5 ml iodine in 5 ml distilled water.

Sample: Polysaccharides (starch).

Procedure

Experiments	Observations	Inferences
• Take 2 ml of sample • Add 1-2 drops of iodine solutions	Blue color develops	The given samples is starch

Note:

1. Polysaccharides (starch) contain two components; amylose and amylopectin. Amylose contains helical structure in which iodine is trapped giving blue color. On heating the helical structure is distorted and trapped iodine is released thereby changing blue colored solution to colorless. On cooling the helical conformation is reformed and the iodine molecule is retrapped.
2. Glycogen is more branched than starch so holds more iodine giving reddish brown color.

Benedict's Test:

Benedict's test is done for differentiating reducing sugars from non-reducing sugars.

Principle

Reducing sugars in alkaline medium undergo tautomerization to produce enediols. Enediols are strong reducing agents and reduce cupric ions of benedict's reagents to cuprous ions which are precipitated in the form of cuprous oxide (red).

Reaction

$$\text{Reducing sugar} \xrightarrow{\text{hot alkaline medium}} \text{1,2 enediol (very reactive reducing agent)}$$

$$\downarrow Cu^{++} \text{ (Benedict reagent)}$$

$$Cu^{+}$$

$$\downarrow \Delta$$

$$Cu_2O \text{ (red ppt)}$$

Reagents

- Copper sulphate: It provides cupric ions.
- Sodium carbonate: It provides alkaline medium.
- Sodium citrate: It stabilizes cupric ions and prevents precipitation as $Cu(OH)_2$.

Samples: Sugar solutions (glucose, fructose, lactose, maltose).

Procedure

Experiments	Observations	Inferences
• Take 5 ml of Benedict's reagents • Add 8 drops of sugars • Mix and heat for 2 minutes	Brick red precipate	The given sugar samples is reducing sugar

Note:

1. Reducing sugars are the sugar containing free aldehyde or ketones group at anomeric carbons.
2. Benedict's test is also known as semi-quantitative test because depending upon the color produced; the concentration of sugars present in the solution can be tentatively estimated.

Color	Sign	Percentage (%)
Blue	-	Nill (absent)
Green	+	Upto 0.5%
Yellow	++	0.5-1%
Orange	+++	1-1.5%
Red	++++	2%
Brick red	++++	>2%

3. Benedicts test is used as screening test in diabetes mellitus (detection of sugar in urine).

Barfoed's Test:

It is done for differentiating reducing monosaccharides from reducing disaccharides.

Principle

Monosaccharides (aldoses and ketoses) can reduce cupric ions in mild acidic condition too and produce red precipitate of Cu_2O.

Reactions

$(CH_3COO)_2\,Cu + 2H_2O \longrightarrow CH_3COOH + Cu(OH)_2$

$Cu(OH)_2 \longrightarrow 2CuO + H_2O$

$2CuO$ + Monosaccharide $\longrightarrow Cu_2O$ (red ppt)

Reagents

- Copper acetate: Provides cupric ions.
- Glacial acetic acid: Provides acidic medium.

Samples: Monosaccharides (glucose, fructose).

Procedure

Experiments	Observations	Inferences
• Take 5 ml of Barfoed's reagents • Add 1 ml of sugar • Heat for 30-45 sec.	Red scum at sides and bottom of test tubes	The given sugar sample is reducing monosaccharide

Note:

1. It is copper reduction test in acidic medium.
2. Monosaccharides react very fast so color develops with 30-45 sec.
3. On prolonged heating disaccharides can also give the test positive.

Seliwanoff's Test:

This test is given positive by keto sugars (C=O group present) and thus helps in distinguishing ketoses from aldoses.

Principle

Ketose sugars on reaction with conc HCl undergo dehydration and produce furfural derivatives which condense with resorsinol to produce cherry red color.

Reactions

$$\text{Fructose} \xrightarrow[H_2O]{\text{conc HCl}} \text{Hydroxymethyl furfural} \xrightarrow{\text{resorcinol}} \text{cherry red colored complex}$$

Reagents: Seliwanoff's reagents (Resorsinol in conc HCl).

Sample: Ketose (fructose).

Experiments	Observations	Inferences
• Take 3 ml of Seliwanoff's reagents • Add 1 ml of sugar • Heat for 30-45 sec.	Cherry red color	The given sugar sample is ketose sugar

Note:

1. Ketoses are more reactive then aldoses in acidic condition, so ketoses readily give this test positive.
2. Heating time is very critical in this test also.
3. Boiling for long time can produce false positive result in case of aldose because it can be converted to ketose by HCl.
4. Sucrose also gives Seliwanoff's test positive because the acidic environment provided by HCl is sufficient to cause hydrolysis of sucrose to produce glucose and fructose. The resuting fructose will then give Seliwanoff's test positive.

Hydrolysis/Inversion Test:

This test is given positive by sucrose (confirmatory test) concentrated acid.

Sucrose is non reducing sugar made up of fructose and glucose. Both fructose and glucose are reducing in nature but during the formation of sucrose the free ends of these sugars are utilised in the formation of α1, β2 glycosidic linkage. So sucrose is non reducing sugar.

Principle

Sucrose before hydrolysis gives Benedict test negative but when sucrose is heated with concentrated acid like HCl, it is hydrolyzed to produce corresponding monosaccharides (glucose and fructose). The sugars being reducing in nature give Benedict test positive.

Sucrose can give Seliwanoff test positive before hydrolysis as acidic environment present in the reagent is sufficient to break α1, β2 glycosidic linkage. The resulting fructose will then give the Seliwanoff test positive.

Reagents

- Concentrated HCl.
- 10% Na_2CO_3.
- Benedict reagent.
- Seliwanoff reagent.

Sample: Sucrose

Procedure

- Take 5 ml of sucrose.
- Add 8-10 drops of conc. HCl.
- Mix and heat for 3 minutes.
- Divide the solution into two parts.
- Neutralize one part by adding 10% Na_2CO_3.
- Perform Benedict test in neutralized part and Seliwanoff test in non-neutralized part.

Experiments	**Observations**	**Inferences**
Benedict test • Take 5 ml of Benedict's reagents • Add 8 drops of neutralized solution • Mix and heat for 2 minutes	Brick red ppt.	The given sugar sample is sucrose as it gives benedict test and seliwanoff test positive after hydrolysis
Seliwanoff test • Take 3ml of Seliwanoff's reagents. • Add 1 ml of non neutralized solution and heat for 30-45 sec.	Cherry red color	

Note: Starch is also non-reducing sugar but on acid hydrolysis it yields the monosaccharide units which give Benedict test positive. Process is similar to hydrolysis of sucrose but heating time should be increased to 5 minutes. Neutralization is required as Benedict reaction occurs in alkaline medium.

Osazone Test:

Reducing sugar can be distinguished by phenyl hydrazine test with the formation of characteristic osazone crystals. These crystals have definite structure, precipitation time and melting point which help in their identification.

Principle

When reducing sugars are treated with phenylhydrazine, first phenyhydrazones are formed which on heating reacts further with another molecule of phenyl hydrazine to form osazones. The reaction takes place in first and second carbon of the sugar, so glucose, fructose and mannose lead to the formation of same osazone crystal. These sugars differ in structures at first and second carbon atom, so they form same osazone crystals as the difference is masked by phenyl hydrazine reaction.

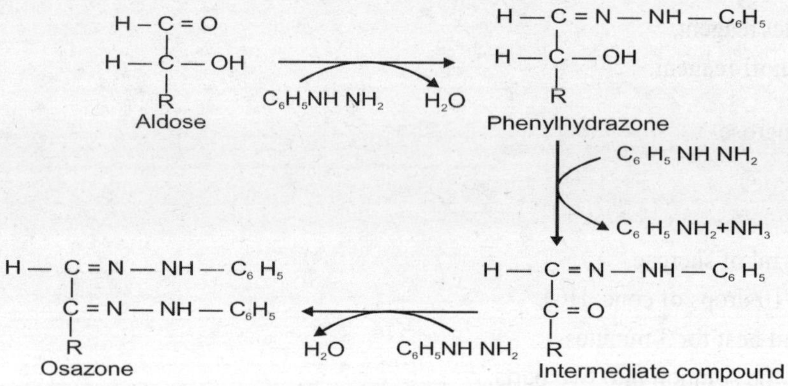

Reagents

- Phenyl hydrazine mixture.
- Glacial acetic acid.

Sample: Glucose, fructose, lactose, maltose.

Procedure

Experiments	Observations	Inferences
• In a clear test tube take 0.5 gm of phenylhydrazine mixture (equal part of phenylhydrazine hydrochloride and anhydrous sodium acetate) • Add 5 ml of sugar solution and 1-2 drops of glacial acetic acid • Mix and keep the test tubes in boiling waterbath for 30 minutes	Osazone crystals Needle shaped Sunflower shaped Cotton ball shaped	The given sugars are: Glucose, Fructose, Mannose Maltose Lactose

Note: Osazones of monosaccharides are insoluble in hot water so they form crystals in hot condition but that of disaccharides are soluble in hot water, so they form osazone crystals on cooling.

Figures showing color reactions of carbohydrates

Test	Negative	Positive
Molisch Violet ring at the interface		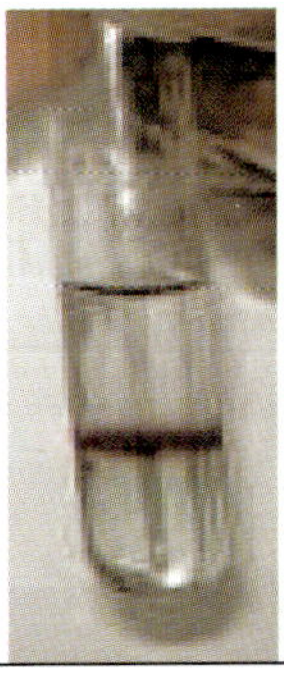
Iodine i. Blue: Starch ii. Violet: Dextrin		i.  ii. 
Benedict Green to brick red color		

Barfoed Red scum at the sides and bottom of test tube	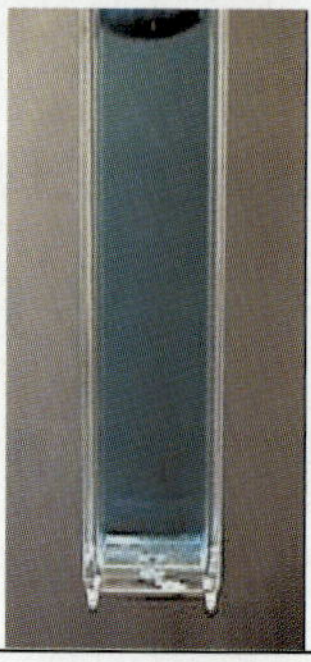	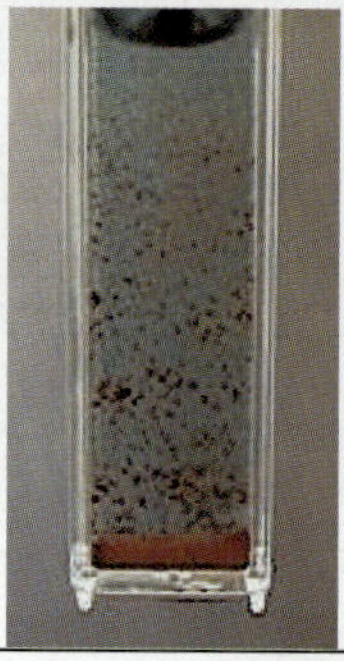 
Seliwanoff Cherry red color	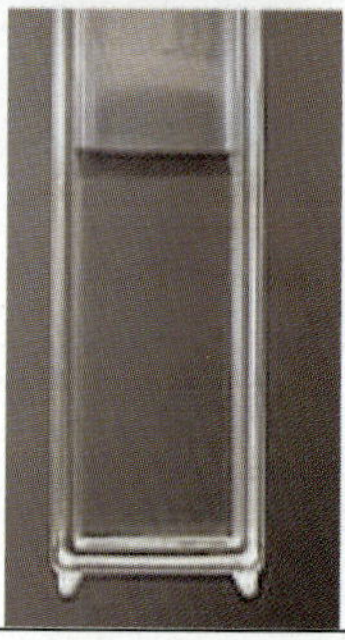	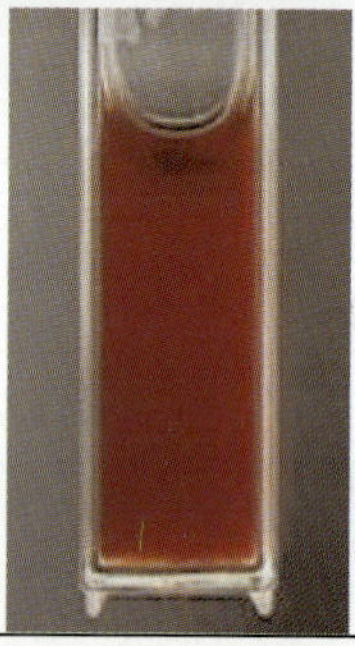
Osazone Needle shaped: Glucose, Fructose, Mannose Sunflower/Petal shaped: Maltose		 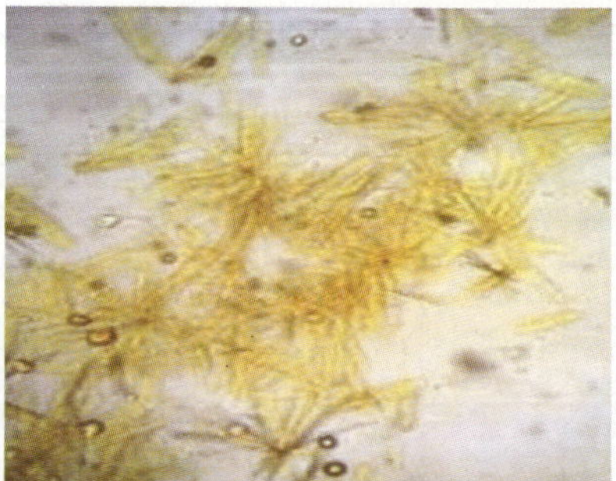

Cotton ball shaped: Lactose		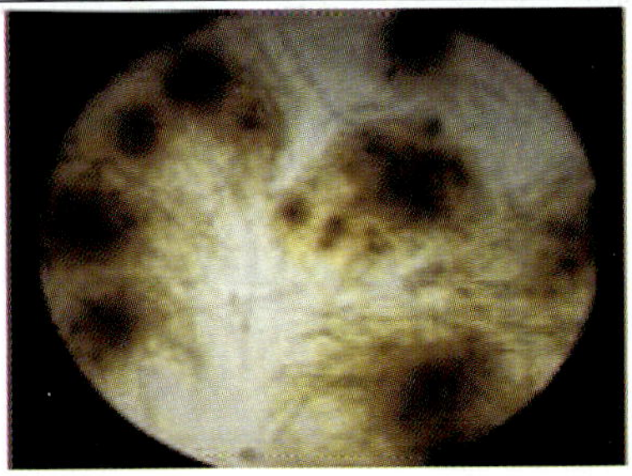
Rhombic plate shaped: Galactose		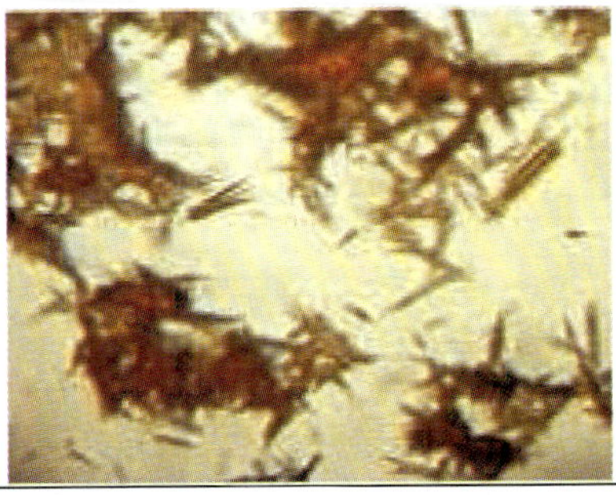

5

IDENTIFICATION SCHEME FOR UNKNOWN CARBOHYDRATE

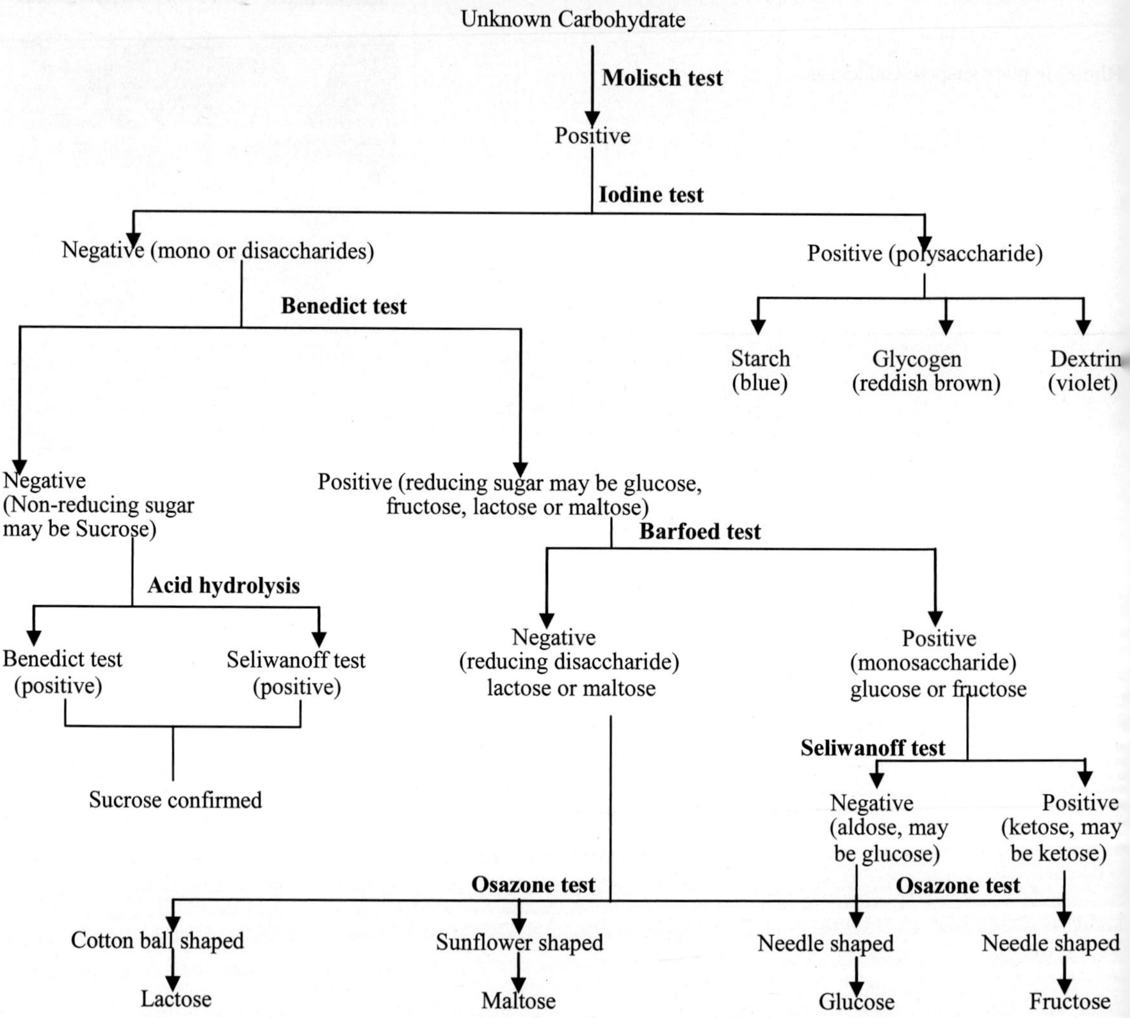

6

PRECIPITATION REACTIONS OF PROTEIN

Proteins are polymers of amino acids joined by peptide bonds. They are made up of carbon, hydrogen, nitrogen and sometimes sulfur and phosphorous. Proteins may be classified as:

1. Simple proteins (made up of amino acids only).
2. Complex/Conjugated proteins (made up of amino acids and other non-protein components called prosthetic groups).

   E.g: Phosphoproteins, nucleoproteins, lipoproteins etc.

3. Derived proteins (formed from native proteins).

They may be primary derived proteins and secondary derived proteins. Primary derived proteins are formed from intramolecular rearrangement. E.g. proteins, metaproteins. Secondary derived proteins are produced from hydrolysis of native proteins. E.g. proteases, peptides, peptones.

Precipitation of Proteins:

- Solubility of proteins depends on: Proportion and distribution of polar and non-polar groups in the molecule.
- Stability of proteins in solution depends on:
  - ➢ Charges on molecule (Similar charges repel each other and prevent their precipitation).
  - ➢ Degree of hydration (shell of water molecules around the particle).
- Polar groups of protein attract water molecules around them to form shell of hydration.

Therefore, proteins can be precipitated by:

- Removing their shell of hydration.
- Neutralization of electric charges.
- Denaturation.
- Bringing them to isoelectric pH.

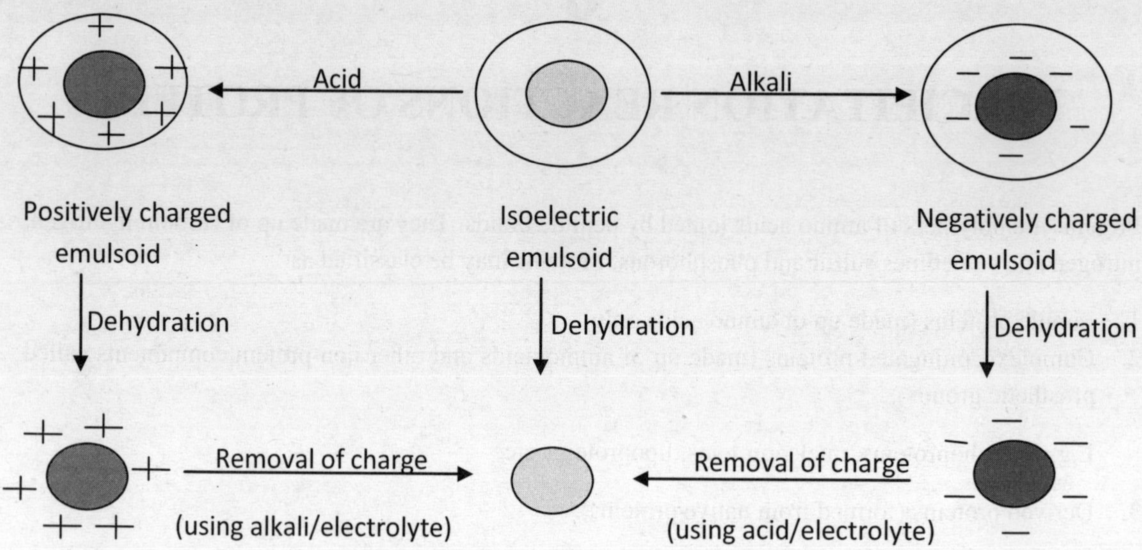

Fig. 6.1: Mechanism of protein precipitation

Precipitation by Salts (Salting Out):

Principle

Proteins are colloidal in nature. Neutral salts like ammonium sulphate or sodium sulphate dehydrate the protein molecule because they have greater affinity for water than colloids. Dehydration causes increased protein-protein interaction leading to molecular aggregation and precipitation.

The amount of salt required to precipitate colloids depends upon:

- Surface area.
- Electric charge.
- Shell of hydration.
- Molecular weight.

Amount of salt required is inversely proportional to the molecular weight and directly proportional to surface area. High molecular weight proteins like globulin, casein, and gelatin have less surface area so less amount of salt is required for precipitation (half saturation). Lower molecular weight proteins like peptone have more surface area so more amount of salt required (full saturation).

Reagents

- Ammonium sulphate crystals (full saturation test).

- Saturated solution of ammonium sulphate (half saturation test).
- Protein solution.

Procedure and Observation

Experiment	Observation	Inference
A. Half saturation test • Take 2 ml of protein solution. • Add 2 ml of saturated solution of ammonium sulphate. • Mix properly and allow standing for sometime.	White precipitate	Given protein solution is precipitated by half saturation with neutral salts.
B. Full saturation test • Take 2 ml of protein solution. • Saturate it with ammonium sulphate crystals. • Allow to stand for some time.	White precipitate	Given protein solution is precipitated by full saturation with neutral salts.

Precipitation with Alkaloid Reagent:

Principle

Proteins are amphoteric in nature (behave as acid in alkaline medium and base in acidic medium). Picric acid, phosphotungstic acid, trichloroacetic acid, sulphosalicylic acid precipitate proteins. They reduce pH of the medium (protein exists in cationic form generally in acidic medium). Negatively charged ions of alkaloids neutralize positive charge on proteins causing precipitation.

Reagents

- Alkaloid reagents: Picric acid, sulphosalicylic acid (20%), trichloroacetic acid.
- Protein solution.

Procedure

Experiment	Observation	Inference
• Take 3 ml of protein solution. • Add sulphosalicylic acid drop by drop. • Mix properly after each addition.	White precipitate	Protein solution is precipitated by alkaloid reagent.
• Take 3 ml of protein solution. • Add picric acid drop by drop. • Mix properly after each addition.	Yellow precipitate	Protein solution is precipitated by alkaloid reagent.
• Take 3 ml of protein solution. • Add trichloroacetic acid drop by drop. • Mix properly after each addition.	White precipitate	Protein solution is precipitated by alkaloid reagent.

Note:

- Use of Tannic acid (alkaloid reagent) in processing of leather is based on protein precipitating effect of tannic acid.
- Sulphosalicylic acid is used in screening the presence of protein in urine and CSF.
- For detection of blood parameters photometrically, serum proteins are removed by adding alkaloid reagents because proteins interfere the estimation.

Precipitation by Heavy Metals:

Principle

At pH more than the isoelectric pH (in alkaline medium), proteins exist as anions (negatively charged ions). Positively charged metal ions complex with protein anion forming metal proteinates which are precipitated. E.g. lead acetate, mercuric nitrate.

Reagents

- Salts of heavy metals like lead acetate (10%), mercuric sulphate (5%).
- Protein solution.

Procedure

Experiment	Observation	Inference
• Take 2 ml of protein solution. • Add 10% lead acetate drop by drop.	White precipitate	Protein solution is precipitated by heavy metals.
• Take 2 ml of protein solution. • Add 5% mercuric nitrate drop by drop.	White precipitate	Protein solution is precipitated by heavy metals.

Note: This is used in treatment of heavy metal poisoning. Salts of heavy metals like lead, mercury, cadmium, zinc etc are toxic as they have the ability to cause precipitation of protein in gastrointestinal tract. Milk or raw egg is used as antidote in heavy metal poisoning because salts of heavy metals disrupt the salt bridges in proteins leading to formation of insoluble metal protein salt that is precipitated. Then an emetic is given to induce vomiting that facilitates removal of metal protein from body.

Precipitation by Organic Solvents (Alcohol, Ether, Acetone):

Principle

Organic solvents reduce shell of hydration and hence the concentration of water required to keep proteins in solution form decreases. They also reduce dielectric constant of medium and cause protein precipitation.

Reagents

- Ethanol, methanol.
- Protein solution.

Procedure

Experiment	Observation	Inference
• Take 1ml of protein solution. • Add 2 ml of ethanol.	White precipitate	Protein solution is precipitated by organic solvents.
• Take 1 ml of protein solution. • Add 2 ml of methanol.	White precipitate	Protein solution is precipitated by organic solvents.

Precipitation by Concentrated Acid:

Principle

Concentrated acid causes denaturation of protein into insoluble metaproteins. These insoluble metaproteins get precipitated. Proteins like albumin, globulin and casein are precipitated on addition of conc HNO_3 whereas derived proteins like gelatin and peptone are not denatured by proteins and do not get precipitated.

Reagents

- Conc HNO_3.
- Protein solution.

Procedure

Experiment	Observation	Inference
Heller's nitric acid test • Take 3 ml of conc. HNO_3 • Add 3 ml of protein solution along the wall of the test tube	White ring at the interface of two liquids	Protein solution is precipitated by concentrated acids.

Heat Coagulation:

Principle

When coagulable proteins are heated, physical, chemical and biological properties are changed due to breaking of certain bonds. As a result of loss of secondary, tertiary and quaternary structures there is disorganization of native protein structure. This is called denaturation of protein. These denatured proteins get easily precipitated. In heat coagulation test, acetic acid is added:

- To provide suitable medium to get maximum precipitate (pH is brought close to isoelectric pH and proteins are precipitated maximally at pI).
- To differentiate between protein and interfering substances mainly phosphates. If precipitate persists and deepens after the addition of acetic acid, it is due to protein. If precipitate disappears after addition of acetic acid, it indicates the presence of phosphates.

Reagents

- 1% acetic acid.
- Protein solution.

Procedure

Experiment	Observation	Inference
• Fill 2/3rd of test tube with protein solution. • Heat the upper 1/3rd portion. • Take lower portion as control. • Add 1% glacial acetic acid. • Coagulum becomes more intensified.	White coagulum forms at upper heated portion that is intensified on addition of acetic acid.	Protein solution is precipitated by heating.

Note: This is widely used for detection of protein (albumin, globulin) in urine in pathological condition.

Isoelectric Precipitation:

Principle

Isoelectric pH (PI) is the pH at which proteins have no net charge. They exist in the form of zwitter ion. Proteins are least soluble at their pI. Hence they can be easily precipitated by bringing the pH of medium in which they are dissolved close to their pI. E.g.: pI of casein: 4.6, pI of albumin: 4.7.

Reagents

- Bromocresol green (BCG).
- 1% acetic acid.
- Protein solution: Casein.

Procedure

Experiment	Observation	Inference
• Take 3 ml of casein solution. • Add 3 drops of BCG. • Add 1% acetic acid drop by drop until light green color is obtained.	Curdy green precipitate	Casein is precipitated at its pI. Green color indicates pH close to 4.6.

Note: The basic principle behind curdling of milk is isolectric precipitation. pH of milk is 6.5-6.7. Lactic acid produced by fermentation process decreases the pH close to pI of casein. Thus casein is precipitated from milk.

- If Biuret test is negative and Nindydrin test is positive, it indicates that only free alpha amino acids are present but no peptide linkages.
- Ninhydrin can be used as staining agent to detect separated amnio acid in chromatography.

Xanthoproteic Test:

It is used to detect presence of aromatic amino acids (tyrosine, phenylalanine and tryptophan).

Principle

Aromatic amino acids contain benzene ring which undergoes nitration when heated with concentrated nitric acid. These nitro derivatives are yellow in color. On addition of alkali the color changes to orange due to ionization.

Reagents

- Concentrated nitric acid.
- 40% NaOH.
- Protein solution.

Procedure

Experiment	Observation	Inference
• Take 3 ml of protein solution • Add 1 ml conc HNO_3 • Mix and heat for 1 minute • Cool under tap water and divide the solution into two parts • To one part add 2 ml of 40% NaOH • Take other part as control	Orange color is developed in test tube in which NaOH is added	The given sample contains aromatic amino acids

Note:

- Phenylalanine give slightly positive test as nitration does not take place under these conditions.
- Skin stains yellow when comes in contact with nitric acid. It is due to nitration of benzene ring of aromatic amino acids in protein.

Millon's Test:

It is a general test for amino acids having hydroxyphenyl group. E.g. tyrosine.

Reagents

- Millon's reagent (10% mercuric sulphate in 10% H_2SO_4).
- 1% $NaNO_2$ (sodium nitrite).

Principle

Sodium nitrite on reaction with sulphuric acid forms nitrous acid. This nitrous acid reacts with phenol group of tyrosine to form nitrated mercuric phenolate ion (red colored complex) on heating.

Procedure

Experiment	Observation	Inference
• Take 1 ml of protein solution. • Add 1 ml of millon's reagent. • Mix and boil for 1 minute. • Cool under tap water. • Add few drops of 1% $NaNO_2$ and warm.	Red colored complex	Presence of hydroxyphenyl group (tyrosine).

Note:

- Heat coagulable proteins like albumin and globulin give red precipitate whereas smaller proteins like peptone give red colored solution.
- Salicylic acid also answers this test as it contain phenolic group.
- Urinary chloride interferes the reaction due to formation of mercuric chloride, so Millon's test cannot be used for detecting tyrosine in urine.

Hopkin-Cole-Adam's Test (Aldehyde Test):

It is the test for indole group containing amino acid, e.g., tryptophan.

Reagents

- Formalin (0.2 % HCHO in water).
- $HgSO_4$ (mercuric sulphate).
- Conc H_2SO_4.

Principle

Sulphuric acid in presence of mercuric sulphate oxidises the indole group which then reacts with formaldehyde to form a violet colored complex.

Procedure

Experiment	Observation	Inference
• Take 3 ml of protein solution. • Add 2 drops of 0.2% formalin. • And a drop of 10% $HgSO_4$. • Mix and add 3 ml of conc H_2SO_4 slowly from the sides of the test tube.	Purple colored ring at the interface of two liquids	Presence of indole group containing amino acid (tryptophan).

Note: This test is given positive by albumin, globulin, casein, peptone but negative by gelatin.

Sakaguchi Test:

It is the test for guanidine group containing amino acid. E.g., arginine.

Reagents

- 5% NaOH.
- Molisch reagent (alpha naphthol).
- Sodium hypobromite (10%).
- Protein solution.

Principle

Guanidino group of arginine in alkaline medium reacts with alpha naphthol and sodium hypobromite to produce bright red colored complex.

Procedure

Experiment	Observation	Inference
• Take 3 ml of protein solution. • Add 1 ml 5% NaOH and 2 drops of Molisch reagent. • Mix and add 10 drops of sodium hypobromite or bromine water.	Bright red color	The given sample contains guanidine group containing amino acid (arginine).

Lead Sulphide Test:

This test is answered by sulphur containing amino acids (cysteine, cystine).

Reagents

- 40% NaOH.
- 10% Lead acetate.
- Protein solution.

Principle

The sulphur present in cysteine and cystine is released as sodium sulphide when heated in strong alkaline medium provided by NaOH. This sodium sulphide then reacts with lead acetate to form black precipitate of lead sulphide.

$$R\text{-}SH + 2NaOH \longrightarrow R\text{-}OH + Na_2S + H_2O$$

$$Na_2S + (CH_3OO)_2Pb \longrightarrow PbS \text{ (black/ brown ppt)} + 2CH_3COONa$$

Procedure

Experiment	Observation	Inference
• Take 2 ml of protein solution. • Add 2 ml of 40% NaOH. • Mix and boil for 1 minute. • Cool and add few drops of 10% lead acetate.	Black or brown ppt	The given sample contains sulphur containing amino acids.

Note:

Methionine, though it is a sulphur containing amino acid answer this test negative because sulphur in methionine is linked with thioether linkage which cannot be easily hydrolyzed by heating with alkali. So sulphur from methione is not released for reaction with lead acetate.

This test is given negative by casein (does not contain cysteine).

Pauly's Test

It is for imidazole group of histidine and phenyl group of tyrosine.

Reagents

- 0.5% sulphanilic acid (0.5% in 2% HCl).
- 0.5% sodium nitrite.
- 10% sodium carbonate.
- Protein solution.

Principle

Diazotised sulphanilic acid reacts with imidazole group of histidine to form red colored complex in alkaline medium. Tyrosine in the same reaction produces orange red color.

Procedure

Experiment	Observation	Inference
• Take 1 ml of 0.5% sulphanilic acid. • Add 1 ml of 0.5 % sodium nitrite solution • Mix and keep for one minute. • Then add 2ml of protein solution and 1 ml of 10% Na_2CO_3	Cherry red color	The given sample contains imidazole group containing amino acid (histidine)

Note:

Tests	Inference
Millon's: Positive, Pauly's test: Positive	Tyrosine
Millon's: Negative, Pauly's test: Positive	Histidine

Neumann's Test:

This test is given positive by phosphoproteins (casein, vitelline).

Reagents

- 5% ammonium molybdate.
- Conc HNO_3.

Principle

On heating with strong NaOH, organic phosphate is released from casein in the form of inorganic phosphate. This phosphate reacts with ammonium molybdate to form canary yellow precipitate of ammonium phosphomolybdate.

Procedure

Experiment	Observation	Inference
• Take 5 ml of protein solution. • Add 0.5 ml of 40% NaOH. • Heat strongly and cool under tap water. • Add 0.5 ml of conc HNO_3 and 1 ml of 5% ammonium molybdate. • Warm gently.	Canary yellow precipitate is formed	The given sample contains phosphoprotein (e.g casein, vitelline).

Figures showing color reactions of proteins

Test	Negative	Positive
Biuret test Violet color		
Ninhydrin test Purple colored ppt (Ruheman's purple)	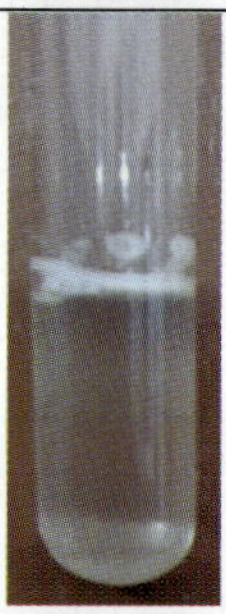	
Xanthoproteic test for aromatic amino acids Yellow in acidic medium Orange in alkaline medium (+)		
Millon's test for tyrosine Red colored ppt/solution	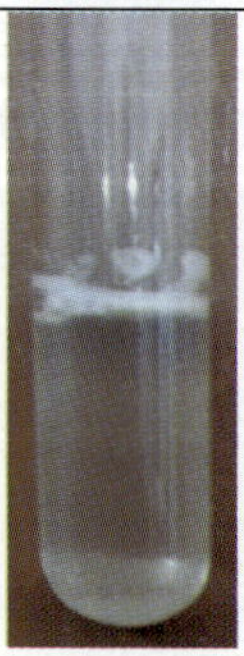	

Aldehyde test for tryptophan Purple ring at the interface		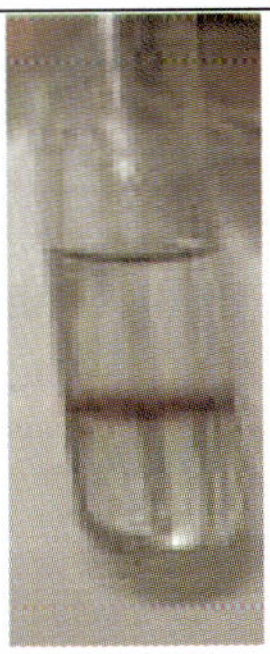
Sakaguchi test for arginine Red colored solution		
Lead sulphide for cysteine **Black/Brown ppt**		
Neumann's test **(phosphoprotein like caesin)**	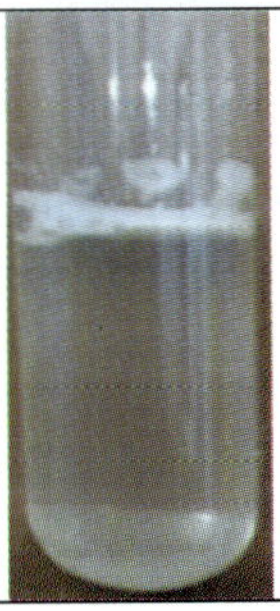	

8

IDENTIFICATION OF UNKNOWN PROTEIN

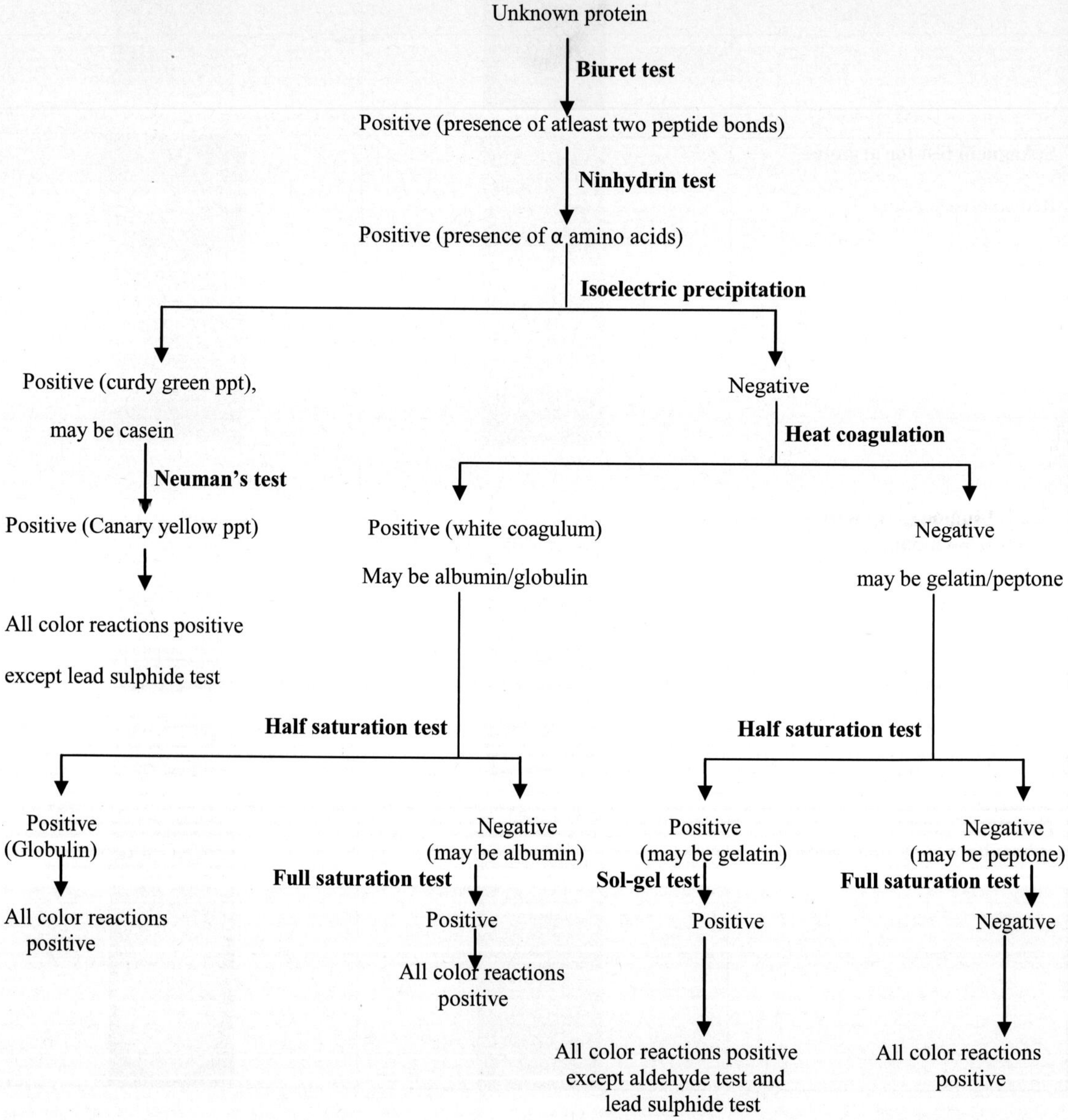

9

GENERAL TESTS FOR LIPIDS

Lipids are heterogenous group of compounds which are relatively insoluble in water but soluble in non-polar solvents like chloroform, ether, benzene etc. Lipids may be classified as:

1. **Simple lipids:** These are esters of fatty acids with glycerol or other higher alcohols. They include Triacylglycerol (esters of fatty acids with glycerol) and waxes (esters of fatty acid with cetyl alcohol).
2. **Compound lipids:** They are esters of fatty acid with alcohol but in addition, they also contain other groups like carbohydrates, proteins, phosphates etc. Depending upon the groups they can be further classified as:

   i. Phospholipids: E.g. lecithin, cephalin.

   ii. Lipoproteins: E.g. HDL, LDL, VLDL.

   iii. Sulfolipids: E.g. sulphated cerebrosides and gangliosides.

   iv. Glycolipids: E.g. cerebrosides, gangliosides.

3. **Derived lipids:** These are derived from lipids or precursor lipids. E.g. cholesterol, steroids.

A. Solubility Tests for Lipids:

Reagents: Ether, chloroform, benzene, carbon tetrachloride

Principle

This test is based on the property of solubility of lipids in organic solvents and insolubility in water. The oil and water separate quickly whereas clear solutions are formed in case of organic solvents showing that oil is insoluble in water and soluble in organic solvents.

Procedure

- Take 3 ml of water, ether, chloroform, benzene and carbon tetrachloride in separate test tubes.
- Add 5 drops of oil in each tube.

Shake vigorously and allow standing for 1-2 minutes.

Observation Table

S.No.	Experiment	Observation	Inference
1	Solubility test with water	Oil and water separate	Oil is insoluble in water
2	Solubility test with ether	It formed clear solution	Oil is soluble in ether
3	Solubility test with chloroform	It formed clear solution	Oil is soluble in chloroform
4	Solubility test with benzene	It formed clear solution	Oil is soluble in benzene
5	Solubility test with Ccl_4	It formed clear solution	Oil is soluble in Ccl_4

B. Grease Spot Test for Lipids:

Reagents: Ether.

Principle

All lipids are greasy in nature. Therefore this test is taken as a group test for lipids.

Observation Table

Experiment	Observation	Inference
Grease spot test • Take 3ml of ether in a test tube. • Dissolve 5 drops of oil in it. • Put a drop of this solution on a filter paper and let it dry. • Ether evaporates leaving a translucent spot on the filter paper indicating the greasy character of the oil.	A translucent spot appears	Lipids are greasy in nature.

C. Emulsification Test with Lipids:

Reagents: Na_2CO_3 (0.5% solution in water), soap solution, bile salt solution (5% in water).

Principle

When oil and water, which are immiscible, are shaken together, the oil is broken up into very tiny droplets which are dispersed in water. This is known as oil in water emulsion. The water molecules, due to their high surface tension, have a tendency to come together and form a separate layer. This is why the oil and water emulsion is unstable. In the presence of substances that lower the surface tension of water, e.g., sodium carbonate, soap, bile salt etc, the tendency of water molecule to coalesce is decreased and the emulsion becomes stable. Since bile salts cause the greatest reduction in surface tension, they are the best emulsifying agent.

Observation Table

S.No.	Experiment	Observation	Inference
1	Emulsification test with:	-	-
a	Water	Oil separates out immediately	Water is non-emulsifying agent
b	Na_2CO_3	Oil separates out a little later	Na_2CO_3 is least emulsifying agent
c	Soap	Emulsification remains for longer period	Soap is more emulsifying agent
d	Bile salt	A permanent emulsion is formed	Bile salts are the most emulsifying agents

Result: Water is non emulsifying agent whereas Na_2CO_3 is least, soap is more and bile salts are the most emulsifying agents.

D. Test for Glycerol:

There are two tests for glycerol:

1. Dunstan's Test

2. Acrolein Test

Dunstan's test is performed with 20% solution of glycerol and Acrolein test is performed with pure glycerol.

1. Dunstan's Test:

Reagents: Borax solution (0.5% solution of sodium borate in water), phenolphthalein indicator.

Principle

Borax hydrolyses into sodium hydroxide and boric acid. As sodium hydroxide is a strong alkali and boric acid a relatively weak acid, the medium becomes alkaline. On adding glycerol, a strong acid, glyceroboric acid is formed and the medium becomes acidic. On heating, glyceroboric acid dissociates into glycerol and boric acid, and the medium becomes alkaline again.

Observation Table

Experiment	Observation	Inference
• Take 3 ml of borax solution. • Add a drop of phenolphthalein. • Add 20% glycerol drop by drop until pink color disappears. • Heat the solution.	Pink color appears (medium is alkaline) Pink color disappears (medium is acidic) Pink color reappears (medium is alkaline again)	Glycerol is present.

2. Acrolein Test:

Reagents: Solid potassium hydrogen sulphate ($KHSO_4$).

Principle:

On heating with potassium hydrogen sulphate ($KHSO_4$), glycerol is dehydrated to form an unsaturated aldehyde, acrolein which has got a characteristic pungent odor.

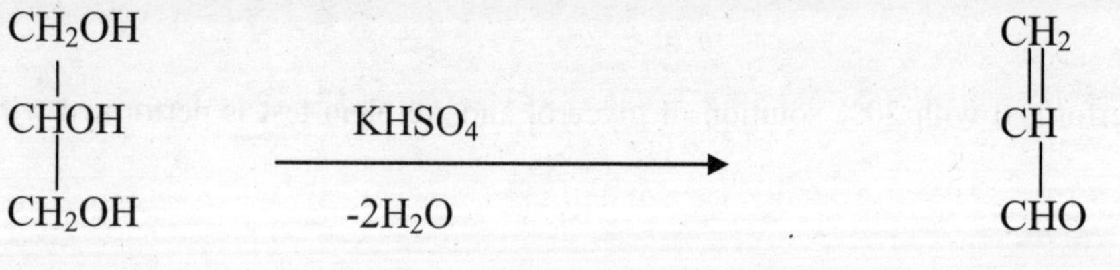

Glycerol Acrolein

Observation Table

Experiment	Observation	Inference
• Take 2 ml of glycerol in a dry test tube. • Add a pinch of $KHSO_4$ and heat.	A pungent smell is observed	Glycerol is present.

E. Tests for Cholesterol:

Some of the tests for cholesterol are Salkowski's test, Libermann-Burchard test etc. These tests are not specific for cholesterol as they are given by all unsaturated sterols.

Sample: 0.5% solution of cholesterol in chloroform

1. Salkowski's test:

Reagent: Conc sulphuric acid

Observation Table

Experiment	Observation	Inference
• Take 2 ml of cholesterol in a dry test tube • Add 2 ml of conc H_2SO_4 by the side of test tube • Allow to stand	Cholesterol and acid layer separate. The cholesterol layer is cherry red color	Cholesterol is present

2. Libermann-Burchard Test:

Reagent: Conc H_2SO_4, acetic anhydride.

Principle

Cholesterol is dehydrated by H_2SO_4 and acetic anhydride to form 3,5-cholestadiene or 2,4-cholestadiene. The cholestadienes combine to form dimers or trimers which react with sulphuric acid to form their sulphonic acid derivatives. The various color appears at different stages of reaction.

Observation Table

Experiment	Observation	Inference
• Take 3 ml of cholesterol solution in a dry test tube. • Add 10 drops of acetic anhydride and 1-2 drops of conc sulphuric acid.	A red color develops followed by blue and finally green	Cholesterol is present.

10

URINE ANALYSIS

Urine is formed in the kidney. The functional unit of kidney is called nephron, where the ultrafiltration of plasma takes place, followed by absorption of most of the water and some of the solutes. The kidneys through nephrons excrete many of the waste products of the body.

Composition of Urine:

1. **Water (95%)**

   - Depends on fluid intake, temperature and physical activity.

2. **Solutes (5%)**

   i. Nitrogenous organic

   ii. Non-nitrogenous organic

   iii. Inorganic

Category	**Compound**	**Normal**
Nitrogenous organic (form bulk of urinary solids)	Urea	15-30 gm/day
	Uric acid	0.5-0.8 gm/day
	Ammonia	0.7-0.8 gm/day
	Creatinine	1-2 gm/day
Non-nitrogenous organic	Presesnt in very small amounts (oxalate, ketones, citric acid, Vitamin C and other water soluble vitamins)	
Inorganic	Sodium	2-4 gm/day
	Potassium	1.5-2 gm/day
	Calcium	0.1-3 gm/day
	Magnesium	0.05-0.2 gm/day
	Phosphate	0.8-1.3 gm/day
	Sulphate	1-1.2 gm/day
	Chlorides	10-15 gm/day

Physical Properties of Urine:

1. **Volume**

   - **Normal:** 1500-2500 ml/day.
   - Depends on fluid intake and loss of fluid by other routes.
   - Output increases when fluid intake is high, salt consumption is increased and in increased intake of beverages like tea, coffee and alcohol.
   - Obligatory volume: 600 ml/day.

2. **Color**
   - **Normal:** Pale yellow.
   - Urine color is due to the presence of pigment urochrome which is a compound of urobilinogen and urobilin. They are final waste product resulting from the breakdown of heme from hemoglobin during the destruction of aging blood cells.
   - Other pigments such as uroerythrin, uroporphyrin and coproporphyrin are present in small amount.

3. **Appearance**

   - **Normal urine:** Clear, transparent without any turbidity or sediment. However on standing urine may become turbid due to precipitation of phosphates. Ammonium ureate may precipitate in alkaline urine and other ureates in acidic urine.

4. **Odor**

   - **Normal urine:** Aromatic odor due to presence of volatile organic acids.
   - On standing ammoniacle odor may be present due to decomposition of urea to ammonia by bacteria.

5. **Specific Gravity:**

   - Urine specific gravity (USG) is a measure of solute concentration in urine and is used to assess the ability of the renal tubules to concentrate or dilute the glomerular filtrate.
   - Normal specific gravity: 1.001 to 1.025 (adults)
1.003 (neonates)
   - It is measured by Urinometer.

6. pH:

- Urinary pH is usually acidic.
- Urine pH is used to classify urine as either a dilute acid or base solution. Depending on the person's acid-base status, the pH of urine may range from 4.5 to 8.
- Average: 6.
- The kidneys play a vital role in the entire process of urine formation and they regulate the acidic levels of urine by absorption of sodium ions and secretion of hydrogen and ammonium ions. The basicity of urine increases when there is absorption of excess alkali or base in the body.
- Excess protein intake produces acidic urine.
- Fruits and vegetables produce alkaline urine.
- It can be tested using pH paper or litmus paper.

CHEMICAL EXAMINATION OF URINE

Test for Organic Constituents:

Test for Ammonia

Principle: Ammonia is alkaline in nature so turns red litmus to blue.

Experiment	Observation	Inference
• Take 5 ml of urine and hold moist red litmus paper over the mouth of test tube	Litmus paper changes from red to blue	Ammonia is present in urine

Test for Urea

A. Hypobromite Test

Principle: Urea is decomposed by sodium hypobromite to form nitrogen and carbondioxide (CO_2). CO_2 is converted to bicarbonate in the presence of alkali.

B. Specific Urease Test

Principle: Soyabean milk contains urease enzyme which hydrolyses urea to ammonia and CO_2.

Experiment	Observation	Inference
A. Specific urease test • Take 3 ml of urine and add 3 drops of phenolphthalein indicator. • Add a pinch of urease powder and mix. • Heat at 45° C or allow to stand for 15 min.	Intense pink color	Urea is hydrolyzed by urease to ammonia which makes the solution alkaline.
B. Hypobromite test • Take 1 ml of urine sample. • Mix with 1 ml of freshly prepared sodium hypobromite solution.	Marked effervescence that disappears quickly	Urea reacts with hypobromite to release nitrogen gas.

Test for Uric Acid (Phosphotungstic Acid Reduction Test)

Principle: Uric acid is a reducing agent. In alkaline condition it reduces phosphotungstic acid to blue colored oxide of tungsten.

Experiment	Observation	Inference
Benedict's uric acid test • Take 2 ml of urine sample. • Add 5 drops of Benedict's uric acid reagent (phosphotungstic acid) and a pinch of anhydrous sodium hydroxide. • Mix properly.	Deep blue colored solution	Uric acid is present which reduces phosphotungstic acid to tungestus salt.

Test for Creatinine (Jaffe's Test)

Principle: Creatinine reacts with picric acid and forms creatinine picrate (orange color) in alkaline condition.

Experiment	Observation	Inference
• Take 2 ml of urine sample. • Mix with 1 ml of saturated picric acid solution. • Add 3 ml of 5% NaOH.	Reddish orange color	Creatinine reacts with picric acid in alkaline medium to form creatinine picrate.

Test for Urobilinogen

Principle: Urobilinogen reacts with paramethyl aminobenzaldehyde of the Ehrlich's reagent to give colored complex.

Experiment	Observation	Inference
• Take 5 ml of urine sample. • Mix properly with 1 ml of Ehrlich reagent and keep for 5 minutes.	Red precipitate	Urobilinogen is present.

Note: This test is positive only for freshly void urine as on standing urobilinogen changes to urobilin which does not give this test positive.

TEST FOR INORGANIC CONSTITUENTS

Test for Calcium

Principle: Calcium present in urine forms white precipitate of calcium oxalate on addition of ammonium oxalate.

Experiment	Observation	Inference
• Take 3 ml of urine and add 1 ml of ammonium oxalate.	White precipitate of calcium oxalate.	Calcium is present in urine.

Test for Phosphate

Principle: Phosphorous of phosphate present in urine is liberated when urine is boiled with conc HNO_3. Phosphorous forms yellow precipitate of ammonium phosphomolybdate when ammonium molybdate is added.

Experiment	Observation	Inference
• Take 3 ml of urine and add 1 ml of conc HNO_3. • Mix, heat and cool it. • Then add 1 ml of ammonium molybdate and heat again.	Yellow precipitate	Phosphate group is present in urine.

Test for Sulphate

Principle: Urine when acidified with HCl forms white precipitate of barium sulphate on addition of barium chloride.

Experiment	Observation	Inference
• Take 3 ml of urine sample. • Then add 1 ml of conc HCl and 1 ml of 10% $BaCl_2$.	White precipitate	Sulphates were precipitated as barium sulphate.

Test for Chloride

Principle: A white precipitate of $AgCl_2$ is formed when acidified urine reacts with $AgNO_3$.

Experiment	Observation	Inference
• Acidify 3 ml of urine sample with 1 ml of conc HNO_3 then add 1 ml of $AgNO_3$.	White precipitate of $AgCl_2$	Urine contains chloride.

11

ABNORMAL COMPONENTS OF URINE

The constituents which are excreted in urine in some diseases or pathological conditions are known as abnormal constituents of urine. These are absent or not present in detectable amounts in normal urine. Example: Glucose, protein, bile pigments, blood etc. 24 hours urine or early morning urine samples are preferred. If there is any delay in analysis then preservatives like thymol should be used.

Abnormal Physical Properties of Urine

Volume

Urine	Volume excreted	Conditions
Polyuria	>2000 ml/day	Excess intake of water and alcohol, diabetes mellitus, diabetes insipidus, intake of diuretics.
Oligouria	<600 ml/day	Fluid deprivation; excess fluid loss due to diarrhoea, vomiting, heat, dehydration, glomerulonephrities, obstruction of urinary tract.
Anuria	<100 ml/day	Renal failure, shock.

Color

Color	Conditions
Pale yellow	Normal
Colorless	Diabetes insipidus, polyuria
Dark yellow	Dehydration (concentrated urine), jaundice (excess of bile pigments).
Dark brown/black on standing	Alkaptonuria (due to excretion of homogentisic acid).
Black	Presence of melanin.
Red	Presence of RBC (hematuria), porphyrins (porphyrias)

Appearance

Appearance	Conditions
Clear/transparent	Normal
Turbid/cloudy	Presence of pus, bacteria, phosphates, carbonates, and ureates
Smoky (brownish gray)	Presence of RBC
Milky	Presence of fat droplets (Chyluria)

Odour

Odor	Conditions
Aromatic	Due to presence of volatile organic acids (Normal)
Ammoniacle	Due to decomposition of urea to ammonia by bacteria
Fruity	Due to presence of acetone (ketonuria occurs in diabetic ketoacidosis, starvation)
Mousy	Phenylketonuria
Mapple syrup	Congenital maple syrup urine disease
Rancid	Tyrosinemia

Specific Gravity

- Normal specific gravity: 1.001 to 1.025.
- Increase in specific gravity of urine is called hypersthenuria (i.e. increased concentration of solutes in the urine). It may be associated with dehydration, diarrhea, emesis, excessive sweating, urinary tract/bladder infection, glycosuria, proteinuria, excess of antidiuretic hormone caused by syndrome of inappropriate antidiuretic hormone etc.
- Decreased specific gravity of urine is called hyposthenuria (i.e. decreased concentration of solutes in urine). It may be associated with renal failure, pyelonephritis, diabetes insipidus, acute tubular necrosis, interstitial nephritis, and excessive fluid intake (e.g., psychogenic polydipsia).

pH

- The pH of urine may range from 4.5 to 8.
- Average: 6.
- A highly alkaline urine occurs in:
  - Metabolic alkalosis in severe vomiting.
  - Respiratory alkalosis due to hypertension.
  - Chronic renal failure.
- A highly acidic urine occurs in:
  - Metabolic acidosis seen in starvation and diabetic ketoacidosis.
  - Respiratory acidosis due to depression of respiratory center.
  - Diarrhea.

CHEMICAL EXAMINATION OF ABNORMAL URINE

Test for Reducing Sugar

- Presence of reducing sugar in urine is called glycosuria.
- Causes: Diabetes mellitus, hyperthyroidism, hyperadrenalism, Pregnancy and lactation (physiological causes).

Test: Benedict Test.

Principle

Reducing sugars in alkaline medium undergo tautomerization to produce enediols. Enediols are strong reducing agents and reduce cupric ions of Benedict's reagents to cuprous ions which is precipitated in the form of cuprous oxide (red).

Procedure

Experiment	Observation	Inference
• Take 5 ml of Benedict's reagent. • Add 8 drops of urine sample. • Mix and heat for 2 minutes.	Greenish to orange colored ppt	Presence of reducing sugar.

Note: Glycosuria may be:

- Renal glycosuria: Blood glucose level is normal but it is excreted in urine due to decrease in renal threshold for glucose (180 mg/dl).
- Alimentary glycosuria: Occurs due to excessive intake of carbohydrates.
- Emotional glycosuria: Occurs in emotional states like anger, fear, stress etc due to hypersecretion of adrenalin that increases blood glucose.

Compounds in urine which are also reducing in nature besides sugar, also give Benedict test, positive. E.g. vitamin C, uric acid, salicylates etc.

Tests for Proteins

- Presence of detectable mount of protein in urine is called proteinuria.
- Causes: nephrotic syndrome, glomerulonephrities, inflammation of lower urinary tract, multiple myeloma (Bence Jones protein excreted).

Test: Sulphosalisylic acid test, Hellers nitric acid test, heat coagulation test.

Procedure

Experiment	Observation	Inference
A. Sulfosalicylic acid test: • Take 2 ml of urine sample and add 2 drops of sulfosalicylic acid.	White turbidity	Protein is present.
B. Heller's nitric acid test: • Take 3 ml of conc HNO_3, Add 3 ml of urine sample slowly from the sides of test tube.	White ring at the interface of two layers	Protein is present.
C. Heat coagulation test: • Take urine sample up to 2/3rd part of the test tube and heat upper 1/3rd part. • Add 2-3 drops of 1% acetic acid.	White coagulum is formed that does not dissolve in acetic acid	Protein is present.

Note:

- Normal amount of protein present in urine is 30-150 mg/day. If excretion is 150-300 mg/day, it is known as microalbuminuria (small amount of albumin is excreted). If the excretion of protein is more than 300 mg/day, it is called proteinuria.
- Bence Jones protein (immunoglobulin) is excreted in multiple myeloma. The feature of this protein is that, this protein precipitates when heated at 40-60 °C. The precipitation disappears at 100 °C and reappears on cooling.
- Acetic acid is added in heat coagulation test.

  ➢ To provide suitable medium to get maximum precipitate (pH is brought close to isoelectric pH and proteins are precipitated maximally at pI).

  ➢ To differentiate between protein and interferring substances mainly phosphates. If precipitate persists and deepens after the addition of acetic acid, it is due to protein. If precipitate disappears after addition of acetic acid, it indicates the presence of phosphates.

Test for Ketone Bodies

Acetone, acetoacetate and β-hydroxybutyrate are called the ketone bodies. Increase in ketone bodies in blood is called ketonemia. Excretion of ketone bodies in urine is called ketonuria.

Principle

Nitroprusside in alkaline solution reacts with keto group of ketone bodies to form a purple colored ring.

Test: Rothra's test.

Procedure

Experiment	Observation	Inference
• Take 5 ml of urine sample. • Dissolve pinch of ammonium sulphate crystals. • Add 2 drops of freshly prepared 5% sodium nitroprusside. • Mix and add liquor ammonia from the side of the test tube.	Purple colored ring at the interace of urine and ammonia layer	Presence of ketone bodies.

Note:

- Beta hydroxybutyrate gives this test negative as it does not contain keto group.
- Normal level of ketone bodies in blood is >1mg/dl. The renal threshold for ketone bodies is 3 mg/dl (until blood ketone level exceeds 3mg/dl, it is not excreted in urine).

Tests for Bile Salts

Bile salts are sodium or potassium salts of bile acid (glycocholic acid and taurocholic acid). Bile salts are normally not excreted in urine. They are excreted in obstructive jaundice.

Test: Hay's sulphur test, Petten koffer's test

Principle of Hay's Sulphur Test: Bile salts lower the surface tension due to which the sulphur powder sinks at the bottom of the test tube.

Principle of Petten Koffer Test: Conc H_2SO_4 converts sucrose to furfurals which then condenses with bile salts to form violet colored complex.

Procedure

Experiment	Observation	Inference
A. Hay's sulphur test: • Take 10 ml of urine sample and sprinkle a pinch of sulphur powder.	Sulphur powder sink to bottom	Bile salts present.
B. Petten Koffer's test: • Mix small amount of sucrose crystals with 5 ml of urine. • Add 5 ml of conc H_2SO_4 along the side of the test tube.	Purple colored ring at the junction of the two layers	Bile salts present.

Note: Alcohol and salicylates can give false positive test.

Tests for Bile Pigments

Bile bigments are breakdown products of heme. E.g. bilirubin, biliverdin. It is excreted in urine in jaundice.

Tests: Fouchet's test, Gmelin test.

Principle of Fouchet test: Bilirubin is adsorbed on the surface of the barium sulphate. It is yellow in color. Fouchet's reagent consists of $FeCl_3$ in trichloroacetic acid which acts as an oxidizing agent and oxidises bilirubin to biliverdin. Biliverdin is greenish in color.

Principle of Gmelin test: Nitric acid oxidizes bilirubin to various colored products like biliverdin (green), bilicyanin (blue), Bilifuscin (red) producing various colored rings in the test tube.

Procedure

Experiment	Observation	Inference
Fouchet's test: • Take 5 ml of urine sample. • Add 3 drops of $MgSO_4$ and 5 drops of $BaCl_2$. Mix properly. • Filter the precipitate and let the residue it dry in oven. • Add few drops of Fouchet's reagent on the residue left in filter paper. • Dry it again in oven.	White ppt obtained Greenish blue color	Formation of $BaSO_4$ Bile pigment present
Gmelin test • Take 5 ml of conc HNO_3. • Slowly add 5 ml of urine solution.	Different colored ring at the point of contact of two liquids	Urine contains bile pigments

Test for Urobilinogen

Principle: Urobilinogen reacts with paradimethyl-aminobenzaldehyde of Ehrlich's reagent to form red colored complex.

Test: Ehrlich's test

Experiment	Observation	Inference
• Mix 5 ml of urine with 1ml of Ehrlich aldehyde reagent (para-dimethylamino-benzaldehyde dissolved in conc HCl) and allow to stand for 5 minutes.	Red color	Presence of urobilinogen.

Test for Blood

- When blood is present in urine, it is called hematuria. Hematuria may be:
  - **Gross hemturia:** Urine appears reddish due to blood (seen in renal stones, malignancies, acute glomerulonephritis etc).
  - **Microscopic hematuria:** Not visible to naked eye (can occur in malignant hypertension, polycystic kidney disease etc).
- Hemoglobinuria is referred to the excretion of hemoglobin through urine due to hemolysis.

Test: Benzidine test.

Principle

In this test peroxidase property of heme is used.

$$H_2O_2 \longrightarrow H_2O + [O]$$

The nascent oxygen reacts with benzidine to form blue colored complex.

Experiment	Observation	Inference
• Take 3 ml of saturated benzidine solution in glacial acetic acid • Add 2 ml of urine sample • Then add 1 ml of H_2O_2	Deep blue or green color	Blood present

Note:

- The blue or green color should develop within 10 minutes. This indicates positive Benzidine test. If the color develops after 10 minutes, it is considered negative as the color production is due to oxidation of benzidine by air.
- Benzidine should be prepared freshly.

Figures showing positive tests for urine sample

Test	Positive
Benedict's test	
Tests for proteins a. Heat coagulation test b. Sulphosalicylic acid test c. Heller's nitric test	 a b c
Rothra's test for ketone bodies	
Hay's sulphur test for bile salt a. Control (sulphur powder donot sink) b. Test (sulphur powder sinks)	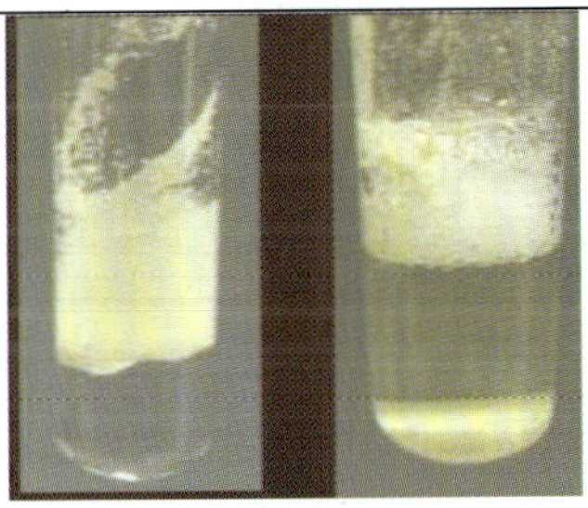 a b
Fouchet's test for bile pigment	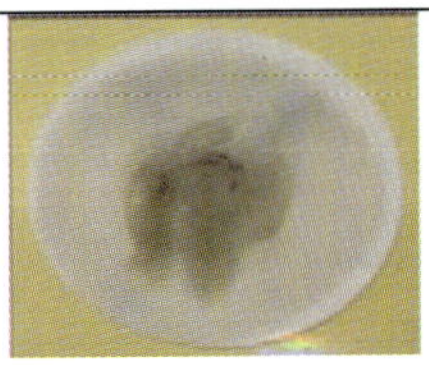

12

PRINCIPLES OF COLORIMETRY

The term colorimetry means measurement of color. In biochemistry, concentration of analytes present in clinical samples can be measured by converting them to colored compounds by suitable reactions. It is very sensitive technique and requires a very small amount of the sample.

Principle: When white light (Incident light I_0) passes through a colored solution, some part of it is absorbed by the coloring substance and the rest is transmitted (emergent light I_e).

Transmittance (T) = Amount of light transmitted through a colored solution

$= (\frac{Ie}{I0})$ [ratio of emergent to incident light]

Absorbance (A)= Amount of light absorbed by colored solution

$= Log(\frac{I0}{Ie})$ [log of ratio of incident to emergent light]

$= Log \frac{1}{T}$

The amount of light absorbed or transmitted depends on two factors:

- Concentration of analyte in the solution.
- Thickness of the solution traversed by light (also called pathlength).

Based on these two factors, there are two laws of colorimetry:

- Beer's Law.
- Lambert's Law.

Beer's Law: If the pathlenght traversed by light is kept constant then absorbance is directly proportional to the concentration (C) of substance in the solution.

A α C

Lambert's Law: If the concentration of substance in the solution is kept constant then absorbance is directly proportional to the pathlength travelled by light (L).

A α L

Beer-Lambert's Law

It is the combination of Beer's law and Lambert's law. The law states that absorbance is directly proportional to the concentration of the chromogen and the thickness of the solution through which the light passes.

$$A \ \alpha \ Ct$$

$$A = KCt$$

If the pathlength (L or t) is kept constant then

$$A \ \alpha \ C$$

$$A = KCt$$

If, Au= the absorbance of unknown solution

Cu= the concentration of substance being measured in the unknown solution.

K= a constant (common to both the solution); and

t= thickness of both the solutions, then

$$Au = K \times Cu \times t$$

$$\text{Or, } \frac{Au}{Cu} = K \times t \ldots\ldots\ldots\ldots\ldots (1)$$

If, As= The absorbance of the standard solution

Cs= The concentration of substance being measured in the standard solution.

$$As = K \times Cs \times t$$

$$\text{Or, } \frac{As}{Cs} = K \times t \ldots\ldots\ldots\ldots\ldots\ldots (2)$$

K= A constant and t= Thickness of both the solutions

From 1 and 2, $\frac{Au}{Cu} = \frac{As}{Cs}$

Therefore,

$$Cu = \frac{Au}{As} \times Cs$$

Type of Solutions Used in Colorimetry

1. Blank: It contains all the solvents and the reagents but not the analyte whose concentration is to be measured.

2. Standard: It is the solution in which the concentration of the analyte to be measured is known.

3. Test: It is the solution in which concentration of analyte to be measured is not known.

Photoelectric Colorimeter

It is the device that measures absorbance of particular wavelength of light by a colored solution. It follows Beer-Lambert's law and is used to measure the concentration of the substance using absorbance.

The basic components of a photoelectric colorimeter are:

1. **Light source:** It is tungsten lamp that provides energy in a visible spectrum (380-750 nm).
2. **Filters:** They are usually made up of colored glass or dyed gelatin. Suitable filter is selected depending upon the color of the solution. Complementary filters are used for determination. It passes light of very narrow range of wavelength.
3. **Cuvette:** It is made up of borosilicate glass and is used for holding the colored solutions.
4. **Photocell or phototube:** It converts light energy into electrical energy.
5. **A measuring device:** It gives reading directly as transmittance or absorbance or both.

Spectrophotometer

It is used when the solution in either colored or colorless. Wavelength of measurement ranges from visible spectrum to near UV range (280-400 nm) and near IR range (750-1000 nm).

- **Light source:** Deuterium halogen lamp.
- **Cuvette:** Made up of quartz.
- **Prism** or diffraction gratings are used in place of filters. This disperses white light into its components and by passing the dispersed light through a narrow slit, a monochromatic light of a specific wavelength can be obtained.
- It is highly sensitive and accurate than colorimeter.

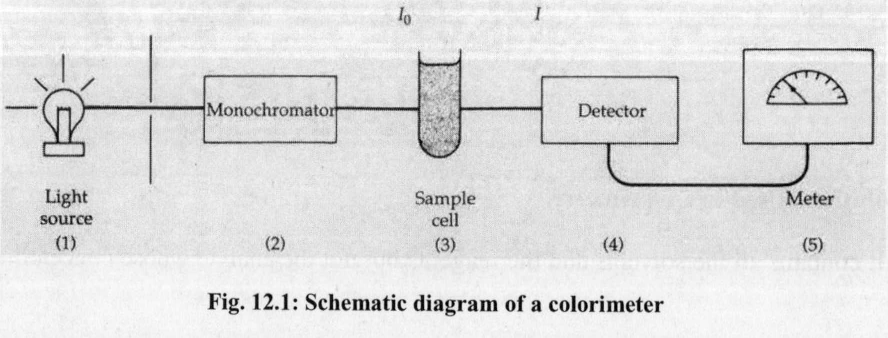

Fig. 12.1: Schematic diagram of a colorimeter

13

VERIFICATION OF BEER-LAMBERT'S LAW

Reagent

$KMnO_4$ (100mg/dl).

Procedure

Mix the solutions as follows and take OD.

S.No.	$KMnO_4$	Distilled water	Concentration	OD at 520 nm
1	50 µl	450 µl	1 mg/dl	
2	100 µl	400 µl	2 mg/dl	
3	150 µl	350 µl	3 mg/dl	
4	200 µl	300 µl	4 mg/dl	
5	250 µl	250 µl	5 mg/dl	

Then plot the graph OD (Absorbance) versus Concentration.

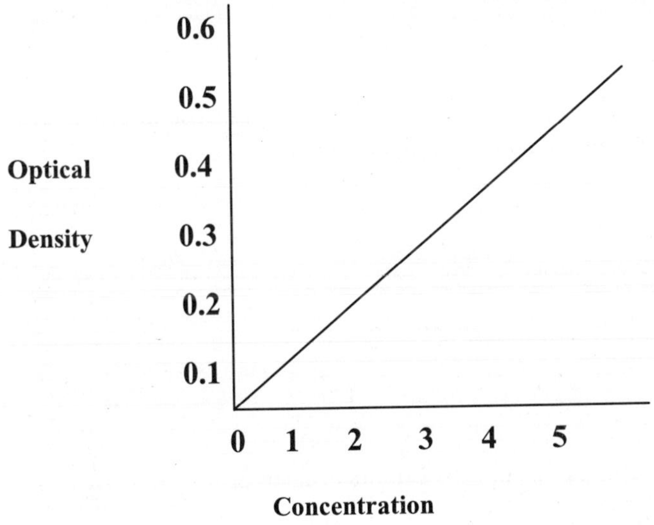

Observation: Straight line passing through the origin.

Result: Beer-Lambert's law is verified.

14

ESTIMATION OF BLOOD GLUCOSE

The term blood sugar means all the sugar (hexoses) present in blood. Glucose is the main hexose present in blood. Thus blood sugar effectively means blood glucose. Glucose is an essential nutrient for brain and RBCs. Level of blood glucose is an important indicator of body function and it is actively maintained in normal range. Abnormalities in its level indicate disturbances in carbohydrate metabolism. Blood glucose estimation is routinely done for diagnosis and follow up of diabetes mellitus patients.

Methods

Broadly two types:

A. Enzymatic Methods:

- Specific methods as the enzymes specifically act on glucose.
- Common methods include:
  1. GODPOD (glucose oxidase and peroxidase)
  2. Hexokinase

B. Chemical Methods:

These are non-relatively specific methods because many substances other than glucose in blood, also cause reaction thereby giving higher results. Common reduction methods include:

1. O-Toluidine method.
2. King Asatoor method.
3. Nelson Somogyi method.
4. Folin Wu method.

Collection of Blood Sample

Samples for blood sugar estimation include:

1. Fasting sample: Sample collected after overnight fast or 10-12 hours of fasting.

2. Post prandial sample: After fasting for 10-12 hours, patient is given carbohydrate meal containing starch and sugar (approximately 100 gm). Then the sample is collected after 2 hours.

3. Random sample: Sample collected at any time in a day irrespective of ingestion of carbohydrate.

Blood is collected in a vial or tube containing potassium oxalate/sodium fluoride mixture (20 mg/5ml) in the ratio of 3:1. Potassium oxalate acts as an anticoagulant. Fluoride inhibits enolase enzyme and hence inhibits the glycolysis. The glucose content will remain unchanged for 2-3 days at 4°C. If fluoride is not added then cells will utilize the glucose and low value may be obtained.

O Toluidine Method

Principle

Glucose reacts with o-toluidine in an acidic medium which is provided by glacial acetic acid to give a blue-green colored complex on heating. The complex is called N-glycosylamine which is measured at 620 nm. The intensity of the color is proportional to the concentration of glucose in the sample.

$$\text{Glucose} + \text{O-toluidine} \xrightarrow[\text{Heat (100°C)}]{} \text{N – glycosylamine derivative}$$

Reagents

- O-toluidine reagent.
- Trichloro acetic acid (10%).
- Standard glucose solution (100 mg/dl=1 mg/ml).

Procedure

A. Preparation of protein free filtrate (PFF) from blood

- Mix 1.5 ml of 10% trichloroacetic acid and 1 ml of serum.
- Centrifuge at 3000 rpm for 5 minutes.
- Collect the supernatant (PFF).

Note: If protein free sample that is prepared directly in distilled water is provided, then the above steps need not to be done. Then take three test tubes and label B (blank), S (standard), T (test).

B. Color development

Reagents	**Blank (B)**	**Standard (S)**	**Test (T)**
Distilled water (D/W)	1 ml	-	-
Standard glucose solution	-	1 ml	-
Test solution/PFF	-	-	1ml
O-toluidine reagent	3 ml	3 ml	3 ml

Mix the contents of each test tubes and incubate in boiling waterbath (100°C) for 10 minutes, cool and take OD or absorbance at 620 nm after setting colorimeter to zero with blank.

Observation

OD of Blank= 0

OD of Standard=

OD of Test=

Calculation:

$$\text{Conc of glucose} = \frac{\text{OD of test}}{\text{OD of standard}} \times \frac{\text{concentration of standard}}{\text{volume of sample /serum}} \times 100$$

$$= \frac{\text{OD of test}}{\text{OD of standard}} \times \frac{1}{1} \times 100$$

$$= \frac{\text{OD of test}}{\text{OD of standard}} \times 100 \text{ mg/dl}$$

=............... mg/dl

King and Asatoor Method

RBCs in blood sample are suspended in isotonic copper sulphate solution to prevent hemolysis and leakage of glutathione. RBCs are precipitated along with proteins by sodium tungstate.

Principle

Glucose when heated in alkaline medium, it is changed into enediol form which reduces copper ions to form the precipitate of cuprous oxide (Cu_2O). The reduction reaction of Cu^{++} to Cu^{+} is coupled with phosphomolybdic acid which is converted to molybdenum blue. This blue colored complex is proportional to the glucose level. It is measured at 610 nm.

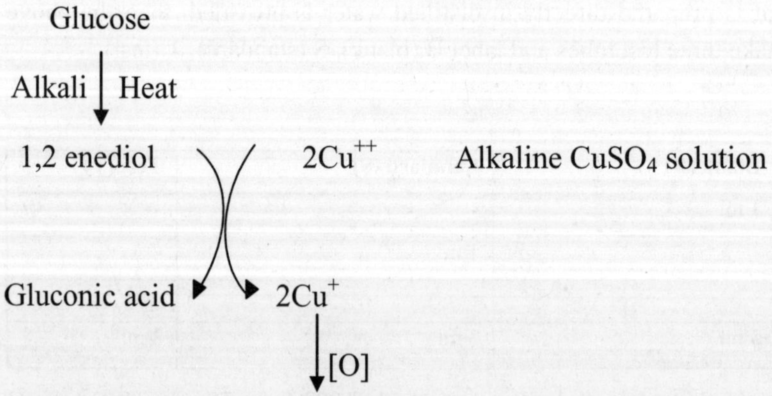

Reagents

- Isotonic $CuSO_4$.
- Sodium tungstate solution (10%).
- Phosphomolybdic acid.
- Standard (10 mg% or 0.1 mg/ml).

Procedure

A. Preparation of protein free filtrate (PFF)

- Take 7.4 ml of isotonic $CuSO_4$, 0.2 ml of whole blood and 0.4 ml of sodium tungstate solution (10%) in a centrifuge tube.
- Mix and leave for 5 minutes.
- Centrifuge.
- The supernatant obtained is PFF.

Note: If protein free sample that is prepared directly in distilled water is provided, then the above steps need not to be done.

B. Reduction of cupric ions

- Take three test tubes and label standard (S), test (T) and blank (B).

Reagents	Blank (B)	Standard (S)	Test (T)
Distilled water (D/W)	2 ml	1.5 ml	-
Standard glucose solution	-	0.5 ml	-
Test solution/PFF	-	-	2 ml
Alkaline $CuSO_4$	2 ml	2 ml	2 ml

- Mix the content in each tube and plug them with cotton.
- Heat in boiling water bath for 10 min.
- Remove the tubes and cool.
- Add 3 ml of phosphomolybdic acid in each tubes.
- Mix and take OD (optical density) at 610 nm after setting the colorimeter to zero by blank.

Observation

OD of Blank=0

OD of Standard=

OD of Test=

Calculation:

1. For serum

1 ml standard glucose= 0.1mg glucose, so 0.5 ml standard= 0.05 mg glucose.

0.5 ml standard is mixed with 1.5 ml distilled water, so dilution is 1: 4 times. Thus amount of standard taken= 0.05 × 4 mg.

$$\text{Conc of glucose} = \frac{\text{OD of test}}{\text{OD of standard}} \times \frac{\text{amount of standard taken}}{\text{volume of sample}} \times 100$$

$$= \frac{\text{OD of test}}{\text{OD of standard}} \times \frac{0.05 \times 4}{0.2} \times 100$$

$$= \frac{\text{OD of test}}{\text{OD of standard}} \times \frac{0.2}{0.2} \times 100 \text{ mg/dl}$$

$$= \frac{\text{OD of test}}{\text{OD of standard}} \times 100 \text{ mg/dl}$$

=............mg/dl

2. For the samples prepared in distilled water

$$\text{Conc of glucose} = \frac{\text{OD of test}}{\text{OD of standard}} \times \frac{\text{amount of standard taken}}{\text{volume of sample}} \times 100$$

$$= \frac{\text{OD of test}}{\text{OD of standard}} \times \frac{0.05 \times 4}{2} \times 100$$

$$= \frac{\text{OD of test}}{\text{OD of standard}} \times 10 \text{ mg/dl}$$

=............ mg/dl

Glucose Oxidase and Peroxidase Method (GOD-POD)

Principle

Glucose oxidase acts on glucose to produce gluconic acid and H_2O_2. Breakdown of H_2O_2 by peroxidase is coupled with the oxidation of phenol. The phenol complexes with 4-amino antipyrine to give red colored complex which is measured at 530 nm.

D- Glucose $\longrightarrow$ Gluconic acid + H_2O_2

H_2O_2 + phenol + 4-aminoantipyrine $\longrightarrow$ Quinoneimine + H_2O

Procedure

Reagents	Blank (B)	Standard (S)	Test (T)
Distilled water (D/W)	10 µl	-	-
Standard glucose solution	-	10 µl	-
Serum	-	-	10 µl
GOD-POD reagent	1 ml	1 ml	1 ml

Incubate the tubes at 37°C for 10 minutes and take OD at 530 nm.

Observation

OD of Blank=0

OD of Standard=

OD of Test=

Calculation:

$$\text{Conc of glucose} = \frac{\text{OD of test}}{\text{OD of standard}} \times \text{conc of standard}$$

=.................... mg/dl

Result

The level of glucose in the given sample is…….. mg/dl.

Interpretation

Normal range of blood sugar:

- Fasting: 60-100 mg/dl.
- Post prandial (PP): 100-140 mg/dl.
- Random: 60-140 mg/dl.

Increase in level of blood glucose is called *Hyperglycemia.* Causes include:

- Diabetes mellitus (most common in which level may rise upto and occasionally above 500 mg/dl).
- Hyperpituitarism, hyperthyroidism.
- Cushing syndrome (hyperadrenalism).
- Pheochromocytoma.
- Pancreatitis.
- A transient rise may occur in emotional states such as anger, anxiety, and fear due to excessive secretion of epinephrine.
- Intracranial diseases like meningitis, tumors and hemorrhage show moderate hyperglycemia.

Decrease in level of blood glucose is called *Hypoglycemia.* Causes include:

- Over administration of insulin in diabetics (most common).
- Tumors of β cells of pancreas.
- Addison's disease (hypoadrenalism).
- Hypothyroidism (Cretinism, myxedma).
- Glycogen storage diseases.

- Starvation.
- Chronic alcoholism.
- Severe exercise.

Diabetes Mellitus

- Most Common cause of hyperglycemia and early detection is necessary due to its complications.
- Fasting sugar > 126 mg/dl and PP sugar > 200 mg/dl confirms diabetes.
- Clinical presentation:
  - **Glycosuria:** Excretion of glucose in urine. Normally glucose is not excreted in urine. When blood glucose level is above the renal threshold (180 mg/dl), it is exceted in urine and this condition is called glycosuria.
  - **Polyuria:** Frequent urination. There is loss of water due to osmotic effect of glucose.
  - **Polyphagia:** Excessive hunger. Due to loss of glucose and inability of cells to utilize glucose, patient feels hungry and eats more.

Complications of Diabetes

A. Acute complications

1. ***Diabetic ketoacidosis:***
   - Results uncontrolled diabetes leading to severe hyperglycemia.
   - Characterised by ketonemia (ketone bodies>1 mg/dl), ketonuria (ketone bodies appear in urine) and acetone breath.

2. ***Hyperglycemic hyperosmolar state:***
   - Due to insulin insufficiency with inadequate fluid intake.
   - Hyperglycemia is very marked (more than 500 mg/dl).
   - No ketosis and ketone bodies usually not seen in urine.
   - Plasma osmolality is increased (330-380 mosmol/L).

B.Chronic Complications

1. ***Microvascular complications (involve fine vessels):***

   - ➢ Retinopathy leading to blindness.
   - ➢ Neuropathy leading to foot ulcers and gangrene.
   - ➢ Nephropathy leading to renal damage.

2. ***Macrovascular complications (involve major vessels):***

   - ➢ Coronary artery disease and MI.
   - ➢ Cerebrovascular disease (stroke).
   - ➢ Peripheral vascular disease.

- In diabetes along with estimation of blood glucose, urine must be tested for reducing sugar, albumin and ketone bodies.

Glucose Tolerance Test

- Glucose tolerance test may be done orally or intravenously.
- The oral test is preferable.
- In the oral test, a blood sample is collected after an overnight fast, and then, glucose dissolved in 250 ml of water is given orally in the dose of 1.5 gm/kg of body weight subject to a maximum of 75 gm.
- Blood samples (post-glucose load) are collected 30, 60, 90 and 120 minutes after the ingestion of glucose.
- Blood glucose concentration is measured in each sample.
- Any urine passed during the period is qualitatively examined for the presence of glucose.

In normal subjects:

- Fasting blood glucose is within the normal range.

- Blood glucose rises to 130-140 mg/dl an hour after ingestion of glucose.
- Blood glucose returns to the fasting level within two hours of glucose ingestion.
- Glucose is not present in any urine sample.

In diabetic subjects:

- Fasting blood glucose is often above normal.
- The rise in blood glucose after glucose ingestion is excessive.
- Blood glucose does not return to the fasting level in two hours.
- One or more urine samples may contain glucose.

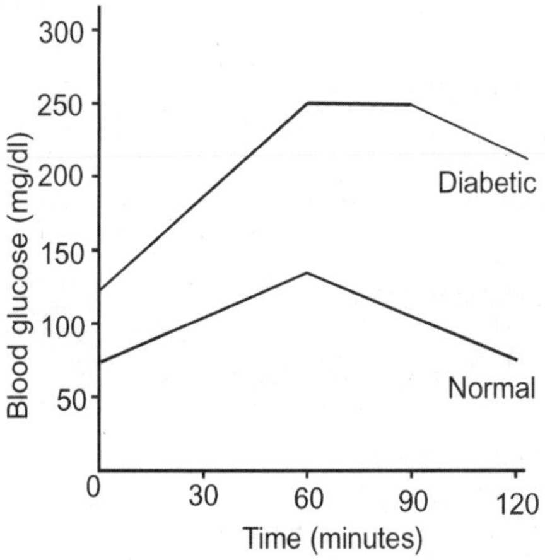

- Once a diagnosis of diabetes mellitus is made and treatment instituted, blood glucose measurements are required to monitor the adequacy of control of diabetes.
- Generally, fasting blood glucose and 120-minute post-prandial (post-meal) blood glucose are measured for monitoring.
- For monitoring long-term control, glycosylated hemoglobin can be measured.

15

ESTIMATION OF SERUM CREATININE BY JAFFE'S METHOD

Creatinine is anhydride of creatine formed by spontaneous, irreversible and non enzymatic reaction (dehydration). It is synthesized in liver (mainly), pancreas and kidneys. It is found in muscles (mainly), liver, brain, testes and blood. It serves as the storage form of energy in skeletal muscles and its level depends on muscle mass. In skeletal muscle, 3/4th of creatine exists as creatine phosphate and 1/4th as free creatine. Approximately 2% of the body creatine is converted to creatinine per day.

$$\text{Glycine} + \text{Arginine} \xrightarrow{\text{amido transferase}} \text{Guanidoacetate} + \text{Ornithine}$$

$$\text{Guanidoacetate} + \text{S-adenosyl methionine} \xrightarrow{\text{methyl transferase}} \text{Creatine} + \text{S- adenosyl homocysteine}$$

$$\text{Creatine} \xrightarrow[\text{ATP} \rightarrow \text{ADP}]{\text{creatine kinase}} \text{Phosphocreatine}$$

Phosphocreatine → (Spontaneous (non enzymatic)) → Creatinine + Pi + H_2O

Reagents

- Saturated picric acid.
- NaOH (10%).
- Working standard of creatinine (2 mg %= 0.02 mg/ml).
- Test creatinine sample.

Principle

Creatinine reacts with saturated picric acid in alkaline medium provided by sodium hydroxide to form orange red colored complex of creatinine picrate (Janovsky complex). The color intensity is directly proportional to the creatinine concentration in the sample and is measured at 520 nm.

$$\text{Creatinine} + \text{Picric acid} + \text{NaOH} \longrightarrow \text{Creatinine picrate (Janovsky complex)}$$

Procedure

A. Preparation of protein free filtrate (PFF) from blood

- Mix 1 ml of distilled water with 1 ml of serum (1:2 dilution).
- Add 1 ml of 10% sodium tungstate and 1 ml of 2/3M H_2SO_4.
- Mix and centrifuge at 3000 rpm for 5 minutes.
- Collect the supernatant (PFF).

Note: If protein free sample that is prepared directly in distilled water is provided, then the above step need not to be done.

B. Color Development

Take three test tubes and label B, S, T.

Reagents	Blank (B)	Standard (S)	Test (T)
Distilled water (D/W)	3 ml	-	-
Standard creatinine solution	-	3 ml	-
Test solution /PFF	-	-	3 ml
10% NaOH	1 ml	1 ml	1 ml
Saturated picric acid	1 ml	1 ml	1 ml

Keep the test tubes at room temperature for 10 minutes and take OD at 520 nm.

Observation

OD of test=0
OD of standard=
Conc. of test=

Calculation:

1. For serum sample

Concentration of creatinine

$$= \frac{\text{OD of test}}{\text{OD of standard}} \times \frac{\text{amount of standard}}{\text{serum}} \times 100 \times \text{dilution factor}$$

$$= \frac{\text{OD of test}}{\text{OD of standard}} \times \frac{0.02}{1} \times 100 \times 2$$

$$= \frac{\text{OD of test}}{\text{OD of standard}} \times 4 \text{ mg/dl}$$

=.......................mg/dl

2. For samples prepared in distilled water

Concentration of creatinine

$$= \frac{\text{OD of test}}{\text{OD of standard}} \times \frac{\text{amount of standard}}{\text{volume of sample}} \times 100$$ [1 ml of standard= 0.02 mg creatinine]

$$= \frac{\text{OD of test}}{\text{OD of standard}} \times \frac{0.02 \text{ X } 3}{3} \times 100$$ [3 ml standard taken =0.02 × 3 mg creatinine]

$$= \frac{\text{OD of test}}{\text{OD of standard}} \times 2$$

=........... mg/dl

Result

The level of creatinine in the given sample is........ mg/dl.

Interpretation

Normal range

- Male: 0.7-1.3 mg/dl.
- Female: 0.6-1.1 mg/dl.
- Higher in males due to more muscle mass.

Creatinine is a tool to assess renal function. Increase in level of serum creatinine occurs in:

- Renal failure.
- Diabetic nephropathy.
- Glomerulonephrities.
- Urinary tract obstruction.
- Dehydration.
- Diseases like gigantism and acromegaly.
- False high values in diabetic ketoacidosis may be seen as Jaffe's reaction is interferred by ketone bodies and glucose.

Decrease in level of serum creatinine occurs in:

- Muscle dystrophy.
- Muscle atrophy.
- Elderly person with small stature.

Estimation of Urinary Creatinine

- Dilute urine to 1:100 if the urine volume is 1-2 litres per day.
- Urine volume <1 L/day: Greater dilution.
- Urine volume >2 L/day: Lesser dilution.
- Creatinine estimation: Jaffe's method and multiply the result by dilution factor.

Limitations of Jaffe's Method

1. Lacks specificity because only 80% of the color developed is due to serum creatinine and rest 20% is due to other substances like glucose, ketone bodies, protein, urea, uric acid etc. which react with alkaline picrate and contribute to color giving rise to higher results. In case of urine color contribution by non-creatinine chromogen is 5%.
2. This test is sensitive to variables like pH, temperature etc.

Clearance

Defined as the volume of plasma (in ml) which is cleared of that amount substance which is excreted by the kidneys in 1 minute.

- Clearance =UV/P where
  - U= Concentration of substance in urine.
  - P= Concentration of substance in plasma.
  - V= Volume of urine (ml/min).

Types of Clearance Tests

- Urea clearance test.

- Inulin clearance test.
- Creatinine clearance test.

Among these, creatinine clearance test is preferred because:

- It is a endogenous constituent of blood.
- Freely filtered and no external infusion required.
- Inulin though ideal substance to determine GFR (freely filtered by glomeruli and neither reabsorbed or secreted by tubules) it is not preferred as it has to be supplied exogenously.
- Not reabsorbed by tubules in large amount as in case of urea (40% of filtered urea is reabsorbed by tubules, so urea clearance is less than GFR).
- Not affected by dietary intake of protein (urea clearance affected by protein intake in diet).

Creatinine Clearance Test

- Creatinine clearance is the volume of plasma cleared of that amount of creatinine which is excreted by kidneys in 1 minute.
- **Normal:** 95-140 ml/min in males

  85-130 ml/min in females.

  Average is 125 ml/min.
- Creatinine clearance is affected at very early stages of glomerular damage when the serum level of creatinine is likely to be normal.
- Normal urinary excretion of creatinine:
  - Male: 1.5-2 gm/day.
  - Female: 0.8-1.5 gm/day.

16

ESTIMATION OF THE SERUM UREA BY DAM (DIACETYL MONOXIME) METHOD

Urea is the end product of protein and amino acid metabolism and constitutes more than 75% of non-protein nitrogen. The ammonia formed from protein metabolism is converted to urea in the liver. Blood urea level measurement is one of the important kidney function tests.

Normal range: 20-40 mg/dl.

Reagents

- Urea standard: 1 mg/dl (0.01 mg/ml)
- Acid reagent: $FeCl_3$ in orthophosphoric acid
- Coloring reagent:
  - ➢ Sulphuric acid
  - ➢ Diacetyl monoxime (DAM)
  - ➢ Thiosemicarbazide

Principle

Diacetyl monoxime when heated in acidic medium decomposes to form diacetyl and hydroxylamine. Diacetyl condenses with urea to give pink colored complex known as diazine. Thiosemicarbazide is used as the catalyst and $FeCl_3$ is used as to intensify the color. The color intensity is directly proportional to the urea concentration in the sample and is measured at 520 nm.

$$\text{Diacetylmonoxime} \longrightarrow \text{Hydroxylamine + Diacetyl}$$

$$\text{Diacetyl + Urea} \xrightarrow{\triangle} \text{Diazine (pink colored complex)}$$

Procedure

Take three test tubes and label B, S, T.

Reagents	Blank (B)	Standard (S)	Test (T)
Distilled water (D/W)	1 ml	-	-
Standard urea solution	-	1 ml	-
Test solution (1:100)	-	-	1 ml
Coloring reagent	2 ml	2 ml	2 ml
Acid reagent	2 ml	2 ml	2 ml

Mix the contents of each test tubes well and keep in boiling waterbath for 20 minutes. Take OD at 520nm.

Observation:

- OD of blank= 0
- OD of test =
- OD of standard =

Calculation:

Sample is diluted first by 1:100

Concentration of urea

$$= \frac{\text{OD of test}}{\text{OD of standard}} \times \frac{\text{amount of standard}}{\text{volume of serum}} \times 100 \times \text{dilution factor}$$

$$= \frac{\text{OD of test}}{\text{OD of standard}} \times \frac{0.01}{1} \times 100 \times 100$$

$$= \frac{\text{OD of test}}{\text{OD of standard}} \times 100$$

=.........................mg/dl

Result

- The concentration of urea in the given sample is ...... mg/dl.

Interpretation

- Elevation of serum urea is known as *uremia.*

Causes of Uremia

Pre-renal Causes

- ➢ Severe vomiting.
- ➢ Prolonged diarrhea.
- ➢ Salt and water depletion.
- ➢ Hemorrhage and shock.
- ➢ Burns.

Renal Causes

- Renal failure
- Acute glomerulonephrities

Post renal causes

- ➢ Prostate enlargement
- ➢ Obstruction of urinary tract
- ➢ Urianary lithiasis
- ➢ Tumor of bladder affecting urine flow

  - Low levels of blood urea is seen in the conditions like

- ➢ Liver failure
- ➢ Liver cirrhosis

BUN (Blood Urea Nitrogen)

$$BUN = \frac{Serum\ urea}{2.14}$$

Estimation of Urea in Urine

- Daily excretion and urinary urea concentration is very high.
- Urine is diluted with distilled water (1:20).
- Urea is then estimated by DAM method.
- Normal range of urinary urea: 20-35 gm/day.

Urea Clearance

- Measure of GFR.
- Volume of plasma which is cleared of that amount of urea that is excreted by kidneys in 1 minute→ urea clearance.
- Depends on urine output.
- When urine output is 2 ml/min or more→ maximal urea clearance.
- When urine output is less than 2 ml/min→ standard urea clearance.

Maximal urea clearance = $\frac{UV}{P}$

Standard urea clearance= $\frac{U\sqrt{V}}{P}$

where,

U= Concentration of substance in urine.

P= Concentration of substance in plasma.

V= Volume of urine (ml/min)

- Normal maximal urea clearance= 75 ml/min.
- Normal standard urea clearance= 54 ml/min .

17

ESTIMATION OF SERUM URIC ACID BY CARAWAY'S METHOD

Uric acid is formed from the breakdown of nucleic acids and is an end product of purine catabolism. Formation of uric acid is mainly endogenous; however, some amount is also formed from purine containing compounds in diet. Liver is the principle organ for uric acid synthesis. Uric acid is transported by the plasma from the liver to the kidney, where it is filtered and excreted in urine. Some of uric acid is excreted into the GI tract. Chemically uric acid is 2,6,8- trihydroxy purine. It acts like a dibasic acid and can form mono and disodium salts depending upon the pH. The salts are deposited in the joints causing arthritis (gout). At pH of 5.75 and below it exist in unionized (acidic) form i.e as uric acid. At pH above 5.75 it is present in ionized form i.e. salt form as monosodium urate. Uric acid is about 10 times less soluble than monosodium urate.

Reagents

- Sodium tungstate (10%).
- Sulphuric acid 2/3 N.
- Sodium carbonate (10%).
- Phosphotungstic acid.
- Uric acid standard (1 mg/dl= 0.01mg/ml).

Principle

Uric acid is a reducing agent. It reduces phosphotungstic acid in presence of sodium carbonate in alkaline medium to produce a blue colored complex, called tungestus complex and itself is oxidized to allantoin. The color intensity is directly proportional to the uric acid concentration in the sample and is measured at 700 nm.

$$\text{Uric acid} + \text{Phosphotungstic acid} \xrightarrow[\text{alkaline medium}]{Na_2CO_3} \text{Tungestus complex} + \text{allantoin}$$

Procedure

A. Preparation of protein free filtrate (PFF) from blood

- Serum proteins interfere with estimation of uric acid, so they are removed.
- Mix 1 ml of serum with 7 ml of distilled water, 1ml of 10% sodium tungstate and 1ml of 2/3N H_2SO_4 (sample dilution is 1:10).
- Centrifuge at 3000 rpm for 5 minutes.
- Collect the supernatant (PFF).

Note: If protein free sample that is prepared directly in distilled water is provided, then the above steps need not to be done.

B. Color development

Take 3 test tubes and label Blank (B), Standard (S) and Test (T)

Reagents	Blank (B)	Standard (S)	Test(T)
Distilled water (D/W)	3 ml	-	-
Standard uric acidsolution	-	3 ml	-
Test solution	-	-	3 ml
10% Na2CO3	1 ml	1 ml	1 ml
Phosphotungstic acid	1 ml	1 ml	1ml

Mix well the contents of each test tube and keep at room temperature for 30 minutes. Take OD against blank at 700 nm.

Observation

OD of Blank=0

OD of Standard=

OD of Test=

Calculation

1. For serum sample

Concentration of standard uric acid=1 mg/dl. i.e. 1ml of standard contains 0.01 mg of uric acid. Thus 3 ml will contain $0.01 \times 3 = 0.03$ mg of uric acid.

Serum sample is diluted 1:10 times and 3 ml of it is taken.Therefore,

10ml solution contains 1 ml serum.

3 ml solution contains $\frac{1\times 3}{10}$ml serum= 0.3 ml.

So, Concentration of uric acid

$$= \frac{\text{OD of test}}{\text{OD of standard}} \times \frac{\text{amount of standard}}{\text{volume of serum}} \times 100$$

$$= \frac{\text{OD of test}}{\text{OD of standard}} \times \frac{0.01 \text{ x } 3}{0.3} \times 100$$

$$= \frac{\text{OD of test}}{\text{OD of standard}} \times 10$$

=.......................mg/dl

2. For samples prepared in distilled water

Concentration of urea

$$= \frac{\text{OD of test}}{\text{OD of standard}} \times \frac{\text{amount of standard}}{\text{volume of sample}} \times 100$$ [1ml of standard= 0.01 mg uric acid]

$$= \frac{\text{OD of test}}{\text{OD of standard}} \times \frac{0.01\text{X } 3}{3} \times 100$$ [3ml of standard used=0.01 × 3 mg uric acid]

$$= \frac{\text{OD of test}}{\text{OD of standard}} \times 1$$

=.................... mg/dl

Result

The concentration of uric acid in the given sample is ...... mg/dl.

Interpretation

- Normal range of uric acid:

  3-7mg/dl (male)

  2-5mg/dl (female)

- Urinary excretion: 400-600 mg/day.
- In women range increases after menopause and in third trimester of pregnancy.
- Up to level of 7 mg/dl uric acid exist in salt form and can remain in dissolved state in 100 ml of blood. At levels more than this uric acid precipitates as mono sodium urate in plasma.
- Estimation of uric acid is important for diagnosis and management of gout in which uric acid is elevated. It is characterised by:
  - Arthritis (occurs due to deposition of monosodium urate crystals in joints→tophi).
  - Recurrent renal stones (renal stones are of uric acid).
- **Gout may be:**
  - **Primary gout:** Develops due defect in uric acid metabolism resulting in increase in uric acid levels. Causes may be HGPRT deficiency, PRPP synthatase overactivity, Glucose-6-phosphatose deficiency etc.
  - **Secondary gout:** Results from other diseases in which increased production of uric acid is a by product of the disease, e.g., polycythemia, leukemia etc.

Increase in serum uric acid concentration→ Hyperuricemia.

Besides gout, other causes of increase in uric acid levels: Renal failure, Von Gierk's disease, Hyperparathyroidism, Lactic acidosis (competes with uric acid excretion), Diabetic ketoacidosis, certain drugs like salicylates, diuretics, alcohols, High intake of organ meats rich in nucleic acid may increase uric acid in healthy person.

Decrease in serum uric acid concentration →*Hypouricemia.* It is less common and may occur in: Hepatocellular disease, Wilson's disease, Fanconi's disease. Overdose of allopurinol, ADA deficiency etc.

Enzymatic Method of Uric Acid Estimation (Uricase Method)

$$\text{Uric acid} + O_2 + H_2O \longrightarrow \text{Allantoin} + H_2O_2 + O_2$$

$$H_2O_2 + \text{phenol} + \text{4-aminoantipyrine} \longrightarrow \text{Quinoneimine} + H_2O$$

18

ESTIMATION OF THE SERUM TOTAL PROTEIN BY BIURET METHOD

Various types of proteins are present in blood and they may be hormones, immunoglobulins, enzymes or factors etc. Palsma proteins mostly include albumin, globulin and fibrinogen. serum lackes fibrinogen as it is used in clotting. Albumin: globulin ration (A:G) are the estimates of renal functions and nutritional status. Blood total protein levels are affected by dietary intake. liver synthesis and renal function (significant urinary loss of proteins occurs in renal disorders like nephrotic syndrome).

Principle

Substances containing at least two peptide bonds react with cupric ions (Cu^{2+}) in alkaline medium to form purple colored complex. The complex is formed between cupric ions, nitrogen atom of peptide bond and oxygen of water molecule. Intensity of the color is directly proportional to the concentration of protein in the sample and is measured at 540 nm.

Reagents

- Biuret reagent (KOH, Hydrated $CuSO_4$ and Sodium potassium tartarate).
- Standard protein solution (0.6 gm % Bovine Serum albumin=600 mg/dl or 6 mg/ml).

Procedure

Take 3 test tubes and label Blank (B), Standard (S) and Test (T).

Reagents	Blank	Standard	Test
Distilled water (D/W)	1 ml	-	0.9 ml
Standard protein solution	-	1 ml	-
Test solution	-	-	0.1 ml
Biuret reagent	3 ml	3 ml	3 ml

Mix the contents of each test tubes well and incubate at 37°C for 10 minutes. Take OD at 540 nm.

Observation

OD of blank= 0

OD of test =

OD of standard =

Calculation: [0.1 ml serum mixed with 0.9 ml distilled water. Dilution is 1:10]

$$\text{Conc of total protein} = \frac{\text{OD of test}}{\text{OD of standard}} \times \frac{\text{amount of standard}}{\text{volume of sample}} \times \frac{1}{1000} \times 100$$

$$= \frac{\text{OD of test}}{\text{OD of standard}} \times \frac{6}{1/10} \times \frac{1}{1000} \times 100$$

$$= \frac{\text{OD of test}}{\text{OD of standard}} \times \frac{6}{0.1} \times \frac{1}{1000} \times 100$$

$$= \frac{\text{OD of test}}{\text{OD of standard}} \times 6 \text{ gm/dl}$$

Result

The concentration of total protein in given sample is .... gm/dl.

Normal Range

- Total protein→ 6-8 gm/dl
- Albumin→ 3.5 to 5.5 gm/dl
- Globulin →2 to 3.5 gm/dl
- Fibrinogen→ 0.2 to 0.4 gm/dl

Clinical Interpretation

1. Decrease in level of serum protein→*Hypoproteinemia*. The causes:

- Kwashiorkor.
- Decreased synthesis, e.g. cirrhosis of liver.
- Increased urinary loss, e.g. nephrotic syndrome, glomerulonephritis.

- Decreased intestinal absorption, e.g. chronic diarrhea.
- Increased catabolism, e.g. hyperthyroidism, high fever.
- Increased loss of blood, e.g. injury, burns and accidental cases.
- Hemodilution like in case of diabetes insipidus, compulsive polydypsea (excessive thirst).

2. Increase in level of serum protein→*Hyperproteinemia*. The causes:
   - Chronic infection like TB, COPD.
   - Malaria.
   - Kala azar.
   - Multiple myeloma.
   - Hemoconcentration, e.g., dehydration, excessive sweating.

19

ESTIMATION OF SERUM ALBUMIN

Albumin is the most abundant protein in human serum or plasma. It is a globular protein produced in the liver. Liver produces about 12 gm of albumin per day (25% of hepatic protein synthesis). It is soluble and monomeric.

Functions:

1. Maintains osmotic pressure

- Contributes 80% to osmotic pressure of plasma.
- Decrease in albumin causes decrease in osmotic pressure leading edema formation.
- Edema is generally seen when blood albumin level is below 2 gm/dl.

2. Transport function

- Transports steroid hormones, thyroid hormones, bilirubin, non-esterified fatty acids, metals like calcium, copper etc.

3. Buffering action

- Albumin has maximum buffering capacity because of its high concentration.
- 16 histidine residues present that contributes to buffering action.

4. Nutritional function

- Can be used as source of amino acid by cells, particularly in nutritional deprivation states.

Methods

A. Biuret Method

Principle

All the substances containing at least two peptide bonds react with cupric ions (Cu^{2+}) in alkaline medium to form purple colored complex which can be measured at 540 nm. The color coordination complex is formed between cupric ions, nitrogen atom of peptide bond and oxygen of water molecule. Intensity of the color is directly proportional to the concentration of protein in the sample.

Reagents

- Biuret reagent (KOH, Hydrated $CuSO_4$ and Sodium potassium tartarate).
- Standard protein solution (0.6 gm % Bovine Serum albumin).

Procedure

Add 0.2 ml of serum to 5.8 ml of sodium sulphite. Mix by inversion and allow standing for 5 minutes. Take clear filtrate for albumin estimation.

Take 3 test tubes and label Blank (B), Standard (S) and Test (T).

Reagents	**Blank**	**Standard**	**Test**
Distilled water (D/W)	3 ml	-	-
Standard protein solution	-	3 ml	-
Filtrate	-	-	3 ml
Biuret reagent	3 ml	3 ml	3 ml

Mix the contents of each test tubes well and incubate at 37°C for 10 minutes. Take OD at 540 nm.

Observation:

OD of blank= 0

OD of test =

OD of standard =

Calculation

$$\text{Conc of albumin} = \frac{\text{OD of test}}{\text{OD of standard}} \times \frac{6}{3} \times \frac{6}{0.2} \times \frac{1}{1000} \times 100$$

$$= \frac{\text{OD of test}}{\text{OD of standard}} \times \frac{12}{0.2} \times \frac{1}{10}$$

$$= \frac{\text{OD of test}}{\text{OD of standard}} \times 6 \text{ gm/dl}$$

=........gm/dl

B. BCG Method

Principle

Bromo Cresol Green (BCG) is yellow in undissociated form (pH <4.7) and blue in dissociated form (pH >4.7). At pH 4.1, only a small fraction is present in undissociated form. Albumin at this pH behaves as a cation and has a binding affinity towards BCG. Binding of albumin to BCG at 4.1 pH produces a blue colored complex. The color intensity is directly proportional to the albumin concentration in the serum sample and it is measured at 620 nm.

Reagents

- BCG (Bromo Cresol Green) reagent.
- Standard albumin solution (0.6 gm %).

Procedure

Take 3 test tubes and label Blank (B), Standard (S) and Test (T).

Reagents	Blank	Standard	Test
Distilled water (D/W)	0.2 ml	-	-
Standard protein solution	-	0.2 ml	-
Test solution	-	-	0.2 ml
BCG reagent	5 ml	5 ml	5 ml

Mix the contents of each test tubes well and incubate at room temperature for 10 minutes. Take OD at 620 nm.

Observation

OD of blank= 0

OD of test =

OD of standard =

Calculation

$$\text{Conc of albumin} = \frac{\text{OD of test}}{\text{OD of standard}} \times \text{conc of standard}$$

Result

The concentration of albumin in given sample is .... gm/dl.

Normal Range

Albumin→ 3.5 to 5.5 gm/dl.

Clinical Interpretation

Decrease in level of serum albumin→*Hypoalbuminemia* (very common). The causes are:

- Kwashiorkor:
- Decreased synthesis, e.g. cirrhosis of liver.
- Increased urinary loss, e.g. nephrotic syndrome, glomerulonephritis.
- Decreased intestinal absorption, e.g. chronic diarrhoea.
- Increased catabolism, e.g. hyperthyroidism, high fever.
- Increased loss of blood, e.g. injury, burns and accidental cases.
- Hemodilution like in case of diabetes insipidus, compulsive polydypsea (excessive thirst).
- Infections.
- Cancers.
- Some genetic abnormalities.

Increase in level of serum albumin→*Hyperalbuminemia.* The causes are:

- Chronic or severe dehydration.

20

ESTIMATION OF SERUM CHOLESTEROL

Cholesterol is the most important sterol present in human. Main site of synthesis is liver. Some amount is also synthesized in intestine, adrenal cortex, gonads and skin. It contains tetracyclic compound called cyclopentano perhydrophenanthrene ring. Cholesterol is present in blood in two forms:

- Free form (30%).
- Esterified form (70%).

It is transported in bound form to lipoproteins mainly HDL and LDL.

Functions

- Maintains fluidity of membrane.
- Insulating effect on nerve fibre and helps in nerve conduction.
- Bile acid and bile salt synthesis. Bile salts are important for fat absorption.
- Synthesis of steroid hormones.
- Synthesis of vitamin D.
- Esterification of long chain polyunsaturated fatty acid.

Reagents

- Acetic acid-ferric chloride reagent.
- Conc sulphuric acid.
- Cholesterol standard (200 mg/dl=2 mg/ml or 0.2 mg/0.1 ml).

Methods

A. Zak's Method

Principle

In strong acidic medium, cholesterol molecule first undergo dehydration to form cholesta-3,5-diene. Then it undergoes oxidation and sulphonation in presence of ferric ions as catalyst to produce red colored bischolesta-3,5-diene. The color produced is proportional to the total cholesterol concentration and is measured at 530 nm.

Cholesterol ⟶ Cholesta-3,5-diene

Cholesta-3,5-diene + H_2SO_4 ⟶ Bischolesta-3,5-diene (red colored complex)

Procedure

Take 2 test tubes and Standard (S) and Test (T).

Reagent	Standard (S)	Test (T)
Sample	-	0.1 ml
Standard cholesterol	0.1 ml	-
$FeCl_3$-acetic acid	4.9 ml	4.9 ml

Mix the contents of each test tube and keep at room temperature for 10 minutes. Follow the steps as:

Reagent	Blank (B)	Standard (S)	Test (T)
$FeCl_3$-acetic acid reagent	3 ml	-	-
Aliquote (S)	-	3 ml	-
Aliquote (T)	-	-	3 ml
Conc H_2SO_4	2 ml	2 ml	2 ml

Mix the contents of each test tubes well and incubate at 60°C for 10 minutes. Take OD at 530 nm.

Observation

OD of blank= 0

OD of test =

OD of standard =

Calculation

In case of standard:

- Cholesterol standard=200 mg/dl or 0.2 mg/0.1 ml.
- 0.1 ml contains 0.2 mg cholesterol.
- 0.1 ml of standard is added to 4.9 ml $FeCl_3$.
- Therefore 5 ml (0.1+ 4.9) contains 0.2 mg cholesterol.
- 3 ml of this solution was taken so; 3 ml contains $0.2 \times 3/5$ mg cholesterol.

In case of test

- 0.1 ml of test is added to 4.9 ml $FeCl_3$, i.e., 5 ml contains 0.1 ml sample.
- 3 ml of this solution was taken, so 3 ml contains 0.1 × 3/5 ml.

Now,

$$\text{Conc of cholesterol} = \frac{\text{OD of test}}{\text{OD of standard}} \times \frac{\text{amount of standard}}{\text{volume of sample}} \times 100$$

$$= \frac{\text{OD of test}}{\text{OD of standard}} \times \frac{0.6/5}{0.3/5} \times 100$$

$$= \frac{\text{OD of test}}{\text{OD of standard}} \times 200$$

=..................mg/dl

Result

The concentration of cholesterol in the given sample is ...... mg/dl.

B. Nair's Method

Principle

When serum is treated with absolute alcohol, cholesterol is extracted from serum and simultaneously proteins are also precipitated. The proteins are removed by centrifugation and the supernatant is treated with $FeCl_3$ color reagent to yield a pink color which is measure colorimetrically at 540 nm.

Reagents

- Absolute alcohol.
- Stock $FeCl_3 \cdot 6H_2O$ color reagent (2.5 gm of $FeCl_3 \cdot 6H_2O$ in 100 ml of 85% orthophosphoric acid).
- Working color reagent (8 ml of stock + 92 ml H_2SO_4).
- Standard cholesterol (200 mg/dl).

Procedure

Take two centrifuge tubes and label Standard (S) and test (T). Proceed as follows:

Reagent	Standard	Test
Absolute alcohol	4.9 ml	4.9 ml
Standard cholesterol	0.1 ml	
Serum		0.1 ml

Mix and centrifuge for 3 minutes and then collect the supernanat and perform as follows:

Reagent	Blank	Standard	Test
Distilled water	3 ml	-	-
Test supernatant	-	-	3 ml
Standard supernatant	-	3 ml	-
Working $FeCl_3$ color reagent	3 ml	3 ml	3 ml

Heat at 80 °C for 10 minutes. Cool and take OD at 540 nm.

Observation

OD of blank= 0

OD of test =

OD of standard =

Calculation

In case of standard:

- Cholesterol standard=200 mg/dl or 0.2 mg/0.1 ml.
- 0.1 ml contains 0.2 mg cholesterol.
- 0.1 ml of standard is added to 4.9 ml absolute alcohol.
- Therefore 5 ml (0.1+ 4.9) contains 0.2 mg cholesterol.
- 3 ml of this solution was taken so; 3 ml contains 0.2 × 3/5 mg cholesterol.

In case of test:

- 0.1ml of test is added to 4.9 ml absolute alcohol, i.e. 5 ml contains 0.1 ml sample.
- 3 ml of this solution was taken, so 3 ml contains 0.1×3/5 ml.

Now,

$$\text{Conc of cholesterol} = \frac{\text{OD of test}}{\text{OD of standard}} \times \frac{\text{amount of standard}}{\text{volume of sample}} \times 100$$

$$= \frac{\text{OD of test}}{\text{OD of standard}} \times \frac{0.6/5}{0.3/5} \times 100$$

$$= \frac{\text{OD of test}}{\text{OD of standard}} \times 200$$

=................mg/dl

Result

The concentration of cholesterol in the given sample is ...... mg/dl.

Normal Range:

- Total cholesterol→150-200 mg/dl.
- Triglyceride→40-150 mg/dl.
- LDL cholesterol→60-140 mg/dl.
- HDL cholesterol→35-65 mg/dl.
- VLDL cholesterol→20-40 mg/dl.

Interpretation

Estimation of serum cholesterol is important because increase in the level of cholesterol is the main factor for atherosclerosis, myocardial infarction, cerebral hemorrhage and thrombosis.

Increase in serum cholesterol concentration→ *Hypercholesteromia.* The causes are:

- Diabetes mellitus.
- Nephrotic syndrome.
- Obstructive jaundice.
- Hypothyroidism.
- Hyperlipoproteinemia type II to type IV.

Decrease in serum cholesterol concentration→ *Hypocholesterolmia.* The causes are:

- Hepatocellular damage.
- Malabsorption syndrome.
- Hyperthyroidism.
- Pernicious anaemia.

Parameters for Assessing the Cardiac Risk

$$\text{Cardiac risk ratio (CRR)} = \frac{\text{Total cholesterol}}{\text{HDL cholesterol}}$$

Male	Female	Risk
<1 – 3.5	<1 – 3.3	Below average
3.5 – 5	3.4 – 4.4	Average
5.1 – 9.6	4.5 – 7.1	Above average (2X)
>9.6	>7.2	Higher (3X)

Friedewald's Equation

LDL cholesterol= Total cholesterol – (HDL+ VLDL)

And, $$\text{VLDL} = \frac{\text{Triglyceride}}{5}$$

Limitations of the Friedewald Equation

The Friedewald equation should not be used under the following circumstances:

- When chylomicrons are present.
- When plasma triglyceride concentration exceeds 400 mg/dL (4.52 mmol/L).
- In patients with dysbetalipoproteinemia (type III hyperlipoproteinemia).

Other methods for cholesterol estimation:

- Sackett's method.
- Kim and Goldberg method.
- Salkowski's method.

Estimation of Cholesterol by Enzymatic Method (CHODPOD Method)

The free cholesterol undergoes oxidation by cholesterol oxidase (CHOD) producing H_2O_2 which gives pink color on reaction with phenol and 4-aminoantipyrene in presence of peroxidase (POD). The intensity of pink colored complex is proportional to cholesterol concentration and is measured at 505 nm.

21

ESTIMATION OF SERUM BILIRUBIN (TOTAL, DIRECT AND INDIRECT) BY MALLOY AND EVELYN METHOD

Bilirubin is the breakdown product of heme part of hem containing compounds like hemoglobin, myoglobin, cytochrome, peroxidase, catalase etc. Hemoglobin is the major source. Hem is catabolized to water insoluble unconjugated bilirubin in reticuloendothelial system (mainly spleen). Unconjugated bilirubin is transported to liver via blood in bound form with albumin. In hepatocytes intracellur proteins like ligandin and Z protein binds with the unconjugated bilirubin. It is then conjugated to glucuronic acid by the enzyme UDP glucuronyl transferase and is converted to mono (20%) and diglucuronides (80%). Conjugated bilirubin is water soluble and is secreted in bile by an active process. This is rate limiting step and is sensitive to the liver damage. Bile is stored in gallbladder and via cystic and common bile duct is released to duodenum, where conjugated bilirubin is reduced by intestinal bacteria to urobilinogen. Most of this is resecreted in bile by enterohepatic circulation and only a small part is secreted in the stool as stercobilinogen and in urine as urobilinogen.

Reagents

1. Diazoblank (15 ml conc HCl in 985 ml distilled water).
2. Diazo reagent (10 ml Diazo A + 0.3 ml Diazo B)
   - Diazo A (1 gm sulphanilic acid in 15 ml conc Hcl and made final volume upto 1 liter by distilled water).
   - Diazo B (0.5 gm $NaNO_2$ in 100 ml distilled water).
3. Standard (10 mg/dl).

Principle

Bilirubin reacts with diazotized sulphanilic acid to form reddish purple colored complex of azobilirubin. It can be estimated by direct and indirect reactions.

Direct reaction: It occurs in aqueous medium and is given only by the water soluble conjugated bilirubin. The reaction is fast and color develops within few minutes.

Indirect reaction: Methanol is added which solubilizes water insoluble unconjugated bilirubin. Thus, it measures total bilirubin (conjugated + unconjugated).

Procedure:

Take four test tubes and label Total test (TT), Total blank (TB), Direct test (DT) and Direct blank (DB).

Reagents	Total test (TT)	Total blank (TB)	Direct test (DT)	Direct blank (DB)
Distilled water	1.8 ml	1.8 ml	1.8 ml	1.8 ml
Test sample	0.2 ml	0.2 ml	0.2 ml	0.2 ml
Diazo blank	-	0.5 ml	-	0.5 ml
Diazo test	0.5 ml	-	0.5 ml	-
Methanol	2.5 ml	2.5 ml	-	-
Distilled water	-	-	0.5 ml	0.5 ml

Take the OD of Direct test and Direct blank after 1 minute at 540 nm. Keep Total test and total blank at room temperature for 30 minutes and take OD at 540 nm.

Observation

OD of Total test (TT)=

OD of Total blank (TB) =

OD of Direct test (DT) =

OD of Direct blank (DB)=

Calculation

$$\text{Conc of total bilirubin} = \frac{\text{OD of TT} - \text{OD of TB}}{\text{OD of standard}} \times \text{conc of standard} \left(\frac{\text{mg}}{\text{dl}}\right)$$

$$\text{Conc of direct bilirubin} = \frac{\text{OD of DT} - \text{OD of DB}}{\text{OD of standard}} \times \text{conc of standard} \left(\frac{\text{mg}}{\text{dl}}\right)$$

Result

The concentrations of total, direct and indirect bilirubin are ...... mg/dl, ... mg/dl and .... mg/dl respectively.

Normal Range

- Total bilirubin→0.2-1 mg/dl.
- Direct/conjugated bilirubin→0.1-0.4 mg/dl.
- Indirect/ unconjugated bilirubin→0.1-0.6 mg/dl.

Interpretation

- Increase in serum bilirubin concentration→ *Hyperbilirubinemia*
- It may be due to:
  - Increased bilirubin synthesis.
  - Decreased hepatic uptake.
  - Decreased hepatic conjugation.
  - Decreased excretion of bilirubin.
- If plasma bilirubin exceeds 1 mg/dl→hyperbilirubinemia.
- Plasma bilirubin between 1-2 mg/dl→latent jaundice.
- When bilirubin level is more than 2 mg/dl, it diffuses in the tissues producing yellowish coloration of sclera, conjunctiva, skin and mucus membrane. This condition is known as jaundice.

Serum total bilirubin is raised in jaundice. Whether the rise is in the conjugated or unconjugated bilirubin or both depends upon the type of jaundice. Jaundice can be classified in several ways. Most commonly it is classified as:

1. Hemolytic/Pre-hepatic Jaundice

- Unconjugated bilirubin is increased. The rate of formation of bilirubin is increased to such an extent that hepatocytes are unable to conjugate and excrete it.
- Causes include:
  - Increased hemolysis (e.g., Sickle cell disease, Glucose-6-Phosphate Dehydrogenase deficiency).
  - Hemolytic disease of newborn.
  - Mismatch blood transfusion.
  - Thalassemia.
  - Pernicious anaemia.

In infants (<1year) blood brain barrier is not fully matured. When blood level of bilirubin is more than 20 mg/dl, the capacity of albumin to bind it is exceeded. Thus, free bilirubin enters the brain and gets deposited resulting in yellowish staining of brain (condition known as kernicterus). It leads to mental retardation, encephalitis and spasticity.

2. Hepatic Jaundice

- Both conjugated and unconjugated bilirubin are increased. The damaged liver cells are unable to conjugate bilirubin so unconjugated fraction is increased. In addition, inflamed and swollen cells may obstruct the biliary canaculi leading to regurgitation of conjugated bilirubin in the circulation. Thus conjugated bilirubin is also raised.
- Causes include:
  - Hepatitis (alcoholic and viral).
  - Liver cirrhosis.
  - Hepatocelluar carcinoma.
  - Toxins and certain drugs.

3. Obstructive/Post Hepatic Jaundice

- Conjugated bilirubin is increased. It is caused by obstruction in biliary channels. Conjugated bilirubin is regurgitated into the systemic circulation. Obstruction may be intrahepatic or extrahepatic.
- Intrahepatic obstruction is caused by biliary cirrhosis, hepatitis etc.
- Extra hepatic obstruction is caused by stones in gall bladder or biliary tract, carcinoma of head of pancreas etc.

Other Forms of Hyperbilirubinemias or Jaundice

A. Congenital Hyperbilirubinemias

1.Criglar-Najjar syndrome 2.Gilbert's syndrome 3.Lucey Driscroll syndrome	Liver cells are unable to take up or conjugate bilirubin. UDP glucuronyl transferase is deficient.
4. Dubin Johnson syndrome 5.Rotor's syndrome	Defective excretion of bile from liver & unconjugated bilirubin is increased.

B. Acquired Hyperbilirubinemias

1. Physiological jaundice

- May occur in infants due to immature hepatic system of conjugation of bilirubin.

- Bilirubin does not increase above 5 mg/dl.
- Disappears by second week of life.

2. Breast milk jaundice

- High level of estrogen derivatives in maternal circulation is excreted in breast milk.
- Inhibits glucuronyl transferase.
- May cause jaundice in newborn.

Note: Pre hepatic, hepatic and post hepatic jaundice also fall in the group of acquired hyper-bilirubinemias.

Clinical Diagnosis of Jaundice

Parameter	Pre-hepatic	Hepatic	Post hepatic
Serum bilirubin	Unconjugated	Both	Conjugated
Urine bilirubin	Absent	Present (↑)	Present (↑)
Urobilinogen	Present	Present (low)	Absent
Stercobilinogen	Present	Present (low)	Absent
Stool color	Dark brown	Clay	Pale
SGPT/SGOT	Normal	Increased	Normal
ALP	Normal	Normal	Increased

22

SERUM TRANSAMINASES

Transaminases or aminotransferases are the group of enzymes that catalyse the transfer of amino group from α amino acids to α keto acids reversibly. This proccss is known as transamination. Examples of transaminases:

- Serum glutamate pyruvate transaminase (SGPT) or alanine transaminase/alanine amino transferase (ALT).
- Serum glutamate oxaloacetate transaminase (SGOT) or aspartate transaminase/aspartate amino transferase (AST).

A. Estimation of ALT Activity

ALT is a cytosolic enzyme synthesized in liver and acts as a marker of liver diseases.

Method: Reitmann and Frenkel method.

Principle

ALT at 37°C and pH 7.4 acts on alanine and α keto glutarate to form pyruvate and L-glutamate. The amount of pyruvate formed is estimated by the reaction of pyruvate with 2,4 dinitrophenyl hydrazine (DNPH). It results in the formation of 2,4 dinitrophenyl hydrazone of pyruvate which is brown colored in alkaline medium and the color intensity is measured at 520 nm.

L-alanine + α ketoglutarate ⟶ pyruvate + L-glutamate

Pyruvate + 2,4-DNP ⟶ 2,4-dinitrophenyl hydrazone of pyruvate (brown colored complex)

Normal Range: Upto 40 IU/L.

1 IU/L: Amount of enzyme activity that catalyses the conversion of 1 micromole of substrate to product per min per litre of sample.

Clinical Interpretation

ALT is a sensitive indicator of parenchymal liver cell diseases.

1. ALT is increased in:

- Acute viral hepatitis (level increases even before the level of serum bilirubin).
- Toxic hepatitis (due to overdose of drugs like paracetamol).
- Tumor of liver.
- Highly increased in hepatic jaundice.
- Mild increase in obstructive jaundice.
- Ischemia of liver.
- Extensive trauma, muscle disease, shock.

2. ALT is decreased in:

- Liver cirrhosis.
- Malabsorption syndrome during abnormal metabolism of pyrimidine.

B. Estimation of AST Activity

Method: Modified Reitmann and Frenkel method

Principle

AST at 37°C and pH 7.4 acts on aspartate and α keto glutarate to form oxaloacetate and L- glutamate. The oxaloacetate formed spontaneously gets decarboxylated to pyruvate. The amount of pyruvate formed is estimated by the reaction of pyruvate with 2,4 dinitrophenyl hydrazine (DNPH). It results in the formation of 2,4 dinitrophenyl hydrazone of pyruvate which is brown colored in alkaline medium and the color intensity is measured at 520 nm.

L-aspartate + α ketoglutarate ⟶ oxaloacetate + L-glutamate

oxaloacetate ⟶ pyruvate + CO_2

Pyruvate + 2,4-DNP ⟶ 2,4-dinitrophenyl hydrazone of pyruvate (brown colored complex)

Normal Range: 5-45 IU/L

Clinical Interpretation

1. AST activity is increased in:

- Myocardial infarction (4-10 times)
- Severe arrhythmias and severe angina
- Acute pancreatitis

- Progressive muscular dystrophy
- Recent brain trauma and its necrosis
- Metastatic tumor of liver
- Pulmonary infection
- Drugs (aspirin, cortisol, erythromycin, nicotinic acid, morphine)

2. AST is decreased in:

- Liver cirrhosis
- Malabsorption syndrome

23

SERUM ALKALINE PHOSPHATASE (ALP)

ALP is present in varieties of tissues. It is highly concentrated in cells lining bile canaliculi of liver, osteoclasts in bone, proximal tubule of kidney, epithelium of intestine and placenta. Each tissue contain a specific isoenzyme of ALP. ALP has a maximum activity at pH 10. It requires zinc ions for its activity and is inhibited by EDTA (removes zinc ions).

Method: Kind and King method.

Principle: At 37^0C ALP reacts with disodium phenyl phosphate buffered at pH 10 to liberate phenol. Phenol then condenses with 4-aminoantipyrene in presence of alkaline oxidizing agent potassium ferricyanide to give red colored complex.

Other Method: Paranitrophenyl Phosphate Method

Principle

Paranitrophenyl phosphate is hydrolyzed by ALP to produce yellow colored complex which is directly proportional to the ALP concentration and is measured at 405 nm.

Normal Range: 4-17 KA units.

ALP activity is expressed in King's Armstrong unit (KAU).

1 KA= The production of 1 mg of phenol by ALP in 100 ml of serum in 15 min.

1 IU of ALP activity=7.1 KAU.

Clinical Significance

ALP activity is increased in:

- Bone diseases (rickets and osteomalacia).
- Bone tumors like osteogenic carcinoma.

- Liver diseases like obstructive jaundice, cirrhosis, extrahepatic biliary obstruction due to stones in bile duct, cancer of head of pancreas, intrahepatic biliary obstruction due to viral or drug induced hepatitis or malignancy.
- Hyperthyroidism.

ALP activity is decreased in:

- Severe anemia.
- Scurvy.
- Kwashiorkor.
- Hypoparathyroidism.

24

ORGAN FUNCTION TESTS

Evaluation of functional status of the organ requires a collection of tests, known as organ profiles or function tests. Some of the organ function tests include:

- Liver function test.
- Kidney function test.
- Thyroid function test.
- Pancreatic function test.
- Gastric function test.

LIVER FUNCTION TESTS

Functions of Liver

1. Excretory function: Excretion of bile pigments, bile salts and cholesterol in bile.
2. Metabolic function: In carbohydrate, lipid, protein, mineral and vitamin metabolisms.
3. Hematological function: Production of plasma proteins and clotting factors like factor V, VII.
4. Storage functions: Storage of glycogen, vitamins A, D and B_{12}, and trace element like iron.
5. Protective functions and detoxification:

➢ Ammonia is detoxified to urea.
➢ Kupffer cells of liver perform phagocytosis.
➢ Metabolism of xenobiotic.

Liver Function Tests are Performed for:

➢ Detection of hepatocellular damage.
➢ Evaluation of the functional status of liver.
➢ Distinguishing between different types of jaundice.

Classification of Liver Functions Tests

There are various tests to assess liver function. They are classified on the basis of:

1. **Excretory Function:** Measurement of bile pigments, bile salts.
2. **Serum Enzymes:** Transaminase (ALT, AST), alkaline phosphatase (ALP), 5'-nucleotidase, LDH isoenzyme.
3. **Synthetic Function:** Prothrombin time, serum albumin.
4. **Metabolic Capacity:** Galactose tolerance and antipyrine clearance.
5. Detoxification.

Test Based on Abnormalities in Pigment Metabolism (Excretory Function)

1. Serum Bilirubin

- Total serum bilirubin: 0.2-1.0 mg/dl
- Unconjugated fraction: 0.1-0.6 mg/dl; and
- Conjugated fraction: 0.1-0.4 mg/dl

Serum bilirubin: 1-2 mg/dl→ latent jaundice

>2 mg/dl → Clinical jaundice

A rise in total serum bilirubin above 2 mg/dl leads to yellow staining of tissues known as jaundice. Jaundice may be:

- Hemolytic (prehepatic): Total and unconjugated bilirubins are raised.
- Hepatocellular (hepatic): Total, unconjugated and conjugated bilirubins are raised.
- Obstructive (post hepatic): Total and conjugated bilirubins are raised.

2. Bile pigments in urine

- Major bile pigment: Bilirubin.
- Normally, it is not present in urine.
- Conjugated bilirubin is water soluble so it is excreted in urine.

3. Urobilinogen in Urine

- Urobilinogen is formed from bilirubin in the intestine.
- Normal urine contains very small amount of urobilinogen.
- Urine urobilinogen in different types of jaundice:

- In hemolytic jaundice: Increased due to increased formation of bilirubin.
- In hepatocellular jaundice: Normal or decreased.
- In obstructive jaundice: Not present in urine.

4. Bile Salts in Urine

- Formed in liver, and excreted in bile.
- In normal subjects and in hemolytic jaundice, there is no obstruction to the flow of bile so bile salts are not present in urine.
- In hepatocellular jaundice, Bile salts appear in urine due to intrahepatic obstruction of biliary canaliculi.
- In obstructive jaundice: Biliary obstruction prevents bile salt excretion in bile causing regurgitation into systemic circulation.

Being water-soluble, they are excreted in urine.

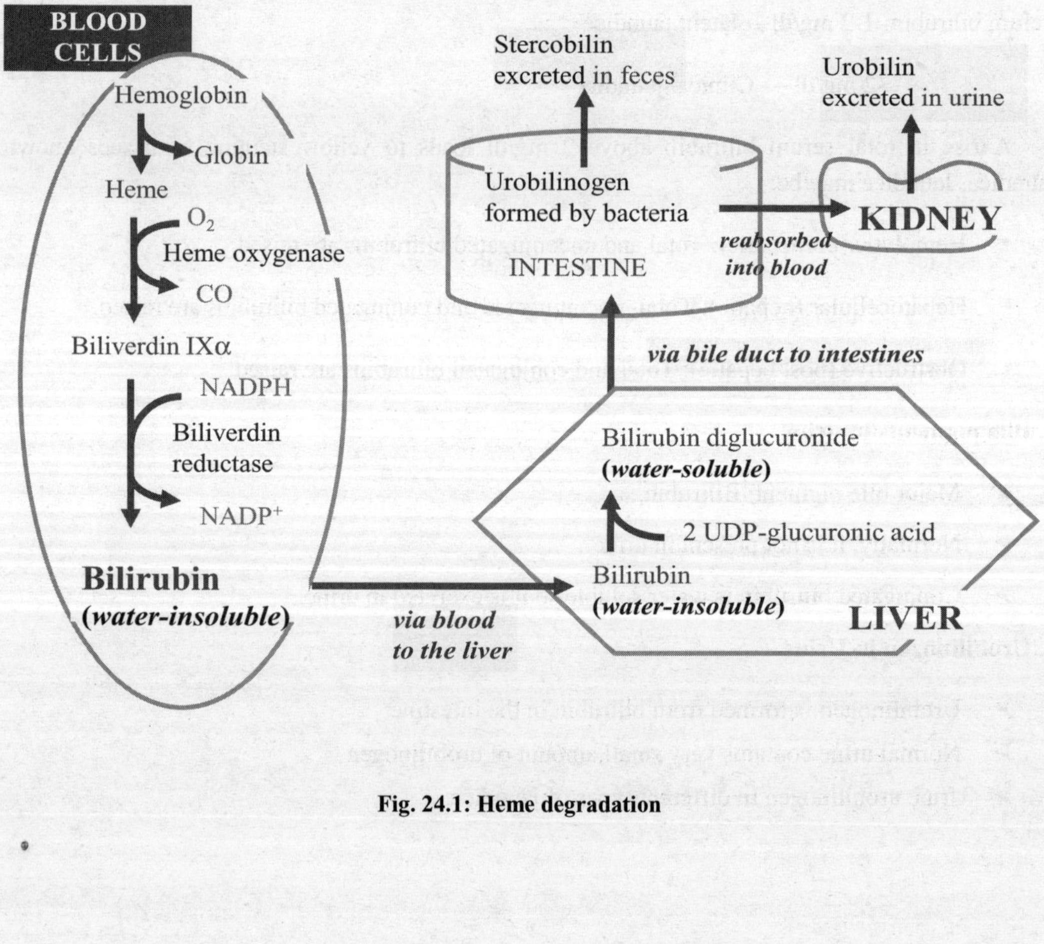

Fig. 24.1: Heme degradation

Comparison of biochemical parameters in different types of jaundice

	Hemolytic jaundice	Hepatocellular jaundice	Obstructive jaundice
Serum bilirubin (total)	↑	↑	↑
Serum bilirubin (unconjugated)	↑	↑	Normal
Serum bilirubin (conjugated)	Normal	↑	↑
Bile pigments in urine	Absent	Present	Present
Urobilinogen in urine	↑	Normal or ↓	Absent
Bile salts in urine	Absent	Present	Present
Faeces color	Dark brown	Pale	Clay
SGOT/SGPT	Normal	Increase	Normal
ALP	Normal	Normal	Increase

Test Based on Carbohydrate Metabolism

1. Galactose Tolerance Test

- Galactose is converted to glucose in liver.
- In this test about, 40 gm of galactose given to the subject after overnight fast.
- Blood galactose is measured after 60 minutes.
- The level above 60 mg/dl indicates impairment of hepatic function.
- In hepatocellular diseases, capacity of the liver to convert galactose into glucose is decreased so blood galactose remains elevated for a long time after ingestion of galactose.

2. Fructose Tolerance Test

- Fructose can be converted to glucose.
- Normal response: No rise in the blood sugar level. The peak should not exceed the fasting level by more than 30 mg % after fructose ingestion.
- In case of infectious hepatitis or parenchymatous liver cells damage, the rise in blood sugar is greater.

TEST BASED ON ABNORMALITIES OF LIPID

Serum Cholesterol

- Normal cholesterol level: 150-200 mg/dl.

- It is increased in obstructive jaundice and decreased in acutc hepatic necrosis and parenchymal liver diseases.
- The total serum cholesterol is usually low and may fall below 100 mg/dl in severe hepatic necrosis.

Fecal Fat: 6 gm/day (normal). It is increased in obstructive jaundice.

Test Based on Amino Acid Metabolism

Nitrogen content of amino acid

↓ Transamination

↓ Deamination

Ammonia in liver (NH_3)

↓

Urea (liver)

Normal NH_3: 40-75 μg/dl.

Increased in advanced cirrhosis, hepatic coma (>200 μg/dl).

Test Based on Synthetic Function

1. Plasma Proteins

Protein	Normal range	Conditions
Albumin	3.5-5.5 gm/dl	• Albumin synthesized only in liver • So level decreases in liver disease
Globulin	2.5-3.5 gm/dl	• Globulin synthesis may be increased, especially in infective diseases • The ratio of albumin: Globulin in serum is decreased or even reversed in liver disease
Fibrinogen	200-400 mg/dl	<100 mg/dl in parenchymal liver damage, cirrhosis

2. Prothrombin Time

- Prothrombin is synthesized in liver.
- Synthesis decreases in hepatocellular disease resulting in prolongation of prothrombin time.
- Prothrombin time is also prolonged in vitamin K deficiency (as vitamin K is required for post-translational modification of newly synthesized prothrombin/pre-prothrombin).
- Parenteral administration of vitamin K restores prothrombin time in vitamin K deficiency but not in hepatocellular disease.

Serum Enzymes

Serum enzymes which are used to ascertain liver function:

- Serum glutamate oxaloacetate transaminase/Aspartate transaminase (SGOT/AST).
- Serum glutamate pyruvate transaminase/Alanine transaminase (SGPT/ALT).
- γ-glutamyl transpeptidase (GGT).
- Alkaline phosphatase (ALP).
- Lactate dehydrogenase (LDH).
- The rise in serum GPT is generally greater than that in GOT in liver disease.
- GOT though reflects damage to the hepatic cells, is less specific for liver disease as it can also be released with heart, muscle and brain disorders.
- Ratios between GPT and GOT are useful in assessing the etiology of liver enzyme abnormalities.
  - ALT or GPT: ASTor GOT= 1.15 (normal).
  - In Viral hepatitis: GPT↑, GOT is normal , GPT/GOT > 1, even more than 2.5.
  - In Chronic hepatitis: GPT↑, GOT ↑GPT/GOT is about 1.
  - In Liver cancer, cirrhosis, Alcohol-induced hepatitis: GPT↑ ,GOT ↑ < 1, about 0.6~0.7.
- GGT is increased in alcoholic hepatitis.
- ALP is mildly elevated in viral hepatitis (necrosis of parenchymal cells).

- A marked elevation in serum ALP occurs in obstructive jaundice (due to irritation of epithelial cells of biliary canaliculi).

- Enzymes like 5' nucleotidase (released by the liver when the liver is injured due to bile duct obstruction or impaired bile flow) and LDH not done routinely.

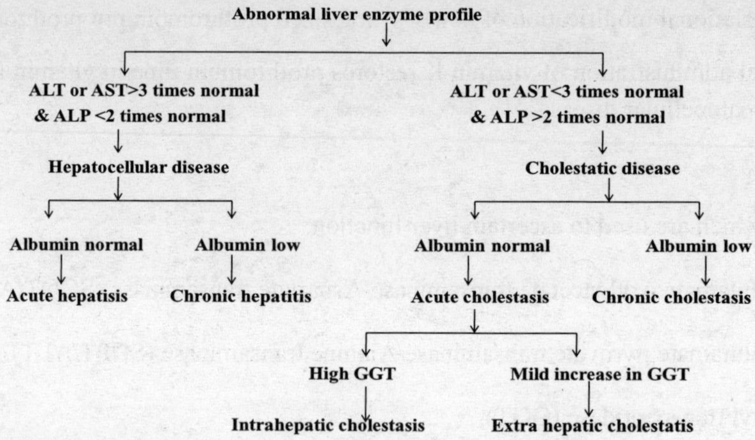

Test Based on Detoxifying Function of Liver

Hippuric Acid Test (Oral or IV)

- This test depends on two factors:
  - Ability of liver cells to provide sufficient glycine.
  - Capacity of liver cells to conjugate it with benzoic acid.
  - Benzoic acid+ Glycine→ Hippuric acid→excreted in urine.

- In case of abnormal liver function, the formation and excretion of hippuric acid is decreased.

- The oral test is done after an overnight fast or 2-3 hours after a light breakfast.

Oral test	Intravenous test
Subject given 6 gm of sodium benzoate in water	Subject asked to void and 1.77 gm sodium benzoate in water injected intravenously
Urine passed for next 4 hours collected	Urine passed for next 1 hour collected
Hippuric acid level measured	Hippuric acid level measured
Excretion of <4gm indicate hepatic impairment	Excretion of <0.8gm indicate hepatic impairment

Test Based on Drug Metabolism:

Antipyerene Breath Test

- Antiphrene is given orally to the patient→ absorbed from intestine→ metabolized in liver by CYP P-450.
- **Normal:** 5-8% of administered dose exhaled.
- **Hepatitis and cirrhosis:** <5% exhalation.

Kidney Function Tests

Kidney function tests aid in the:

- Detection of the presence of an active lesion in the excretory system; and
- Assessment of the functional efficiency of the kidneys.

Kidney function tests may be broadly divided into:

- Tests to assess glomerular function.
- Tests to assess tubular function.

Urine Examination

A routine qualitative examination of urine should be undertaken before quantitative analysis of blood and urine. A careful physical examination and qualitative tests for abnormal constituents of urine, e.g. proteins and blood, may provide useful information, especially about the presence of an active lesion in the urinary tract.

Tests to Assess Glomerular Function:

1. Serum Urea

- Urea is the end product of protein catabolism.
- Synthesized in liver and excreted by kidneys.
- Normal range of serum urea =20-40 mg/dl.
- Impairment of glomerular function decreases glomerular filtration of urea and causes retention in blood. However, serum urea begins to rise only when the glomerular filtration rate decreases below 50% of the normal, so not sensitive indicator.
- The concentration of urea in blood can also be expressed as blood urea nitrogen (BUN) which represents the nitrogen content of urea present in blood.
- Molecular weight of urea is 60 in which the contribution of nitrogen atoms is 28.
- Therefore, BUN=28/60, i.e., nearly 0.47.
- The normal range of BUN is 9-21 mg/dl.
- A rise in nitrogen content of blood is called azotemia.

2. Serum Creatinine

Creatinine is formed from creatine in muscles and is excreted by kidneys in urine. Normal range of serum creatinine is 0.6-1.5 mg/dl in male and 0.7-1.1 mg/dl in females. It is increased in renal failure due to decreased glomerular filtration and, consequently, increased retention of creatinine in blood. Serum creatinine is a more sensitive indicator of glomerular function than serum urea.

Clearance Tests

Clearance is defined as the volume of plasma which is cleared of that amount of substance that is excreted by kidneys in one minute. It can be calculated by:

Clearance = UV/P, where

U (concentration of substance in urine in mg/dl)

V (volume of substance in ml/minute)

S (concentration of substance in serum in mg/dl)

Types of Clearance Tests

- Urea clearance test.
- Creatinine clearance test.
- Inulin clearance test.

1. Urea Clearance Test

Urea is completely filtered by glomeruli but is partially reabsorbed by tubules. Therefore, urea clearance is less than the GFR. Urea clearance is calculated from:

- U (concentration of urea in urine in mg/dl).
- V (volume of urine in ml/minute).
- S (concentration of urea in serum in mg/dl).

If the output of urine is more than 2 ml/minute, then clearance is known as "maximum urea clearance''. Maximum urea clearance (ml/minute) =75 ml/minute in a person having a body surface area of 1.73 square metres.

If the urine output is less than 2 ml/minute, then it is called "standard urea clearance". Standard urea clrearance (ml/minute) = 54 ml/minute in a person having a body surface area of 1.73 square metres.

If the body surface area differs from 1.73 square metres, the calculated clearance value should be corrected by multiplying it by 1.73/body surface area of the subject. Urea clearance below 60% of the normal indicates impaired glomerular function.

2. Creatinine Clearance Test

Creatinine clearance is closer to GFR than is urea clearance since there is little tubular reabsorption of creatinine. The urine passed in the 24-hour period is collected. Blood sample is also collected for creatinine estimation. Then Creatinine clearance is calculated using the formula UV/S. The normal range of creatinine clearance is:

100-120 ml/minute in males.

95-105 ml/minute in female.

Lower than the normal valuesl indicate impairment of glomerular function.

3. Inulin Clearance

Actual/True GFR may be estimated by inulin clearance. Inulin is a polysaccharide of fructose, non-toxic and not metabolized in human beings. It is completely filtered by glomeruli, and is neither secreted nor reabsorbed by tubules. Therefore, its clearance is equal to GFR. Inulin clearance is calculated using the formula UV/S.

Normal inulin clearance is 120-130 ml/minute for a body surface area of 1.73 square metres. Decreased clearance indicates impairment of glomerular function.

Test to Assess Tubular Functions:

1. Concentration Test

When the level of water in the body decreases, tubular reabsorption of water is increased so as to conserve water, and the urine becomes concentrated. In this test, the person is not allowed to take any food or drink after the evening meal. The first three urine samples passed in the morning are collected, and their specific gravity is measured.

In normal individuals, at least one of the urine samples will have a specific gravity of 1.025 or more. If the value remains below 1.025 in all the samples, it indicates impairment of tubular function.

2. Dilution Test

When the water content of body increases, tubular reabsorption of water decreases resulting in the passage of dilute urine. The person is asked to drink 1,200 ml of water within 30 minutes after a overnight fast. Urine samples are collected for four hours at every one hour interval and their specific gravity is measured. In normal subjects, at least one of the samples will have a specific gravity of 1.003 or less. If the specific gravity is above 1.003 in all the specimens, it shows impairment of tubular function.

3. Para-Amino Hippurate Clearance

Para-Amino Hippurate (PAH) is an exogenous compound which is filtered by glomeruli and is also secreted by tubules. PAH is administered by intravenous injection. Urinary and blood PAH concentrations are measured. When present in low amount in blood, the entire PAH present in blood is removed by the kidneys in a single circulation of blood through the kidneys.

PAH clearance is calculated which is equal to the RPF (Renal Plasma Flow). Normal RPF is about 600 ml/minute for a body surface area of 1.73 square metres.

Normal filtration fraction (GFR/RPF) is about 0.2. Filtration fraction decreases in glomerulonephritis.

Thyroid Function Test:

Patients who are likely to be at increased risk of thyroid dysfunction are:

- Patients with other autoimmune diseases (e.g. type 1 diabetes, coeliac disease).

- Patients with dyslipidemia (high cholesterol and/or high triglyceride).
- Chronic cardiac failure, coronary artery disease, arrhythmias, pulse >90/min, hypertension.
- Menstrual disturbance or unexplained infertility.

Thyroid function tests are commonly performed for the diagnosis of hyperthyroidism and hypothyroidism. The following are some of the older tests:

- Basal metabolic rate.
- Serum cholesterol.
- Serum protein-bound iodine.
- Radioactive iodine uptake.

The tests commonly done now are the following:

- Serum T_3 and T_4.
- Serum Free T_3 (FT_3) and Free T_4 (FT_4).
- Serum TSH.
- Thyroperoxidase antibodies (TPO Ab).
- TRH stimulation test.

1. Serum T_4

- The normal range: 5-12 µg/dl.
- Increased in hyperthyroidism and decreased in hypothyroidism.

2. Serum T_3

- The normal range: 0.1-0.2 µg/dl.
- Increased in hyperthyroidism and decreased in hypothyroidism.
- An increase in serum T_3 with normal T_4 is known as T_3 toxicosis.

3. Serum Free T_4 (FT_4) and Free T_3 (FT_3)

In some cases, though there is presence of a thyroid disorder, the level of T_3 and T_4 may be normal. So measurement of FT_4 and FT_3 should be done for proper diagnosis. Normal range for:

FT_4 is 0.8-2 ng/dl

FT_3 is 0.3-0.6 ng/dl

These are increased in hyperthyroidism and decreased in hypothyroidism.

4. Serum TSH

TSH is used as the sole test of thyroid function in most cases and is most sensitive test of thyroid function. It is useful in differentiating between hyperthyroidism and hypothyroidism of thyroid origin and those secondary to pituitary or hypothalamic disease.

In primary hyperthyroidism, serum TSH level is decreased whereas in hyperthyroidism secondary to pituitary or hypothalamic disorder, serum TSH is raised. In primary hypothyroidism, serum TSH is increased but in hypothyroidism secondary to pituitary or hypothalamic disorder, serum TSH is decreased. The normal range of serum TSH is 0.5-5 µU/ml.

5. Thyroperoxidase Antibodies (TPO Ab)

Detection of thyroperoxidase antibodies is useful in diagnosing hypothyroidism caused due to autoimmune disorders like Hashimoto's thyroiditis or postpartum thyroiditis.

6. TRH Stimulation Test

In this test anterior pituitary is stimulated by TRH administration. Then serum T_3, T_4 and TSH are measured. This test is used to distinguish between thyroid disorders of hypothalamic and pituitary origin. In hypothyroidism with low TSH, low T_4 and low T_3, a response to TRH stimulation points to the hypothalamic origin of the disorder.

Comparison of thyroid disorders

Condition	T3	T4	TSH	TRH
1° hypothyroidism	↓	↓	↑	↑
2° hypothyroidism	↓	↓	↓	↑
3° hypothyroidism	↓	↓	↓	↓
1° hyperthyroidism	↑	↑	↓	↓
2° hyperthyroidism	↑	↑	↑	↓
3° hyperthyroidism	↑	↑	↑	↑
Subclinical hypothyroidism	Normal	Normal	↑	-
Subclinical hyperthyroidism	Normal	Normal	↓	-

Pancreatic Function Test:

These tests help in diagnosis of the disorder of exocrine portion of the pancreas. The commonly performed tests include:

- Serum amylase.
- Serum lipase.
- Sweat chloride.

1. Serum Amylase

- Amylase is a carbohydrate digesting enzyme synthesized by pancreas and salivary glands.
- Major percentage of amylase present in circulation is of pancreatic origin.
- It is highly increased in acute pancreatitis due to leakage of the enzyme from the pancreatic cells.
- Mild increase occurs in acute parotitis.

2. Serum Lipase

- It is lipid digesting enzyme synthesized in the pancreas.
- Increased in acute pancreatitis.

3. Sweat Chloride

- It is very useful in the diagnosis of cystic fibrosis (fibrocystic disease of the pancreas).
- Cystic fibrosis is an autosomal recessive disease that affects number of exocrine glands. Highly viscous mucus is secreted which causes clogging of ducts of exocrine gland and bronchioles. In addition, the chloride content of sweat is increased. Diagnosis of cystic fibrosis is done by measuring the chloride concentration in sweat. A concentration above 60 mEq/L is indicative of cystic fibrosis.

Gastric Function Test:

Various test are present for the evaluation of the gastric function.

1. Fractional Meal Test

In this tests first the fasting content of stomach is collected using Ryle's tube. Then the stomach is stimulated by giving a test meal to the patient. After stimulation the gastric contents are again aspirated using Ryle's tube at every 15-minute interval for 2 hours. The collected samples are analyzed for free and total acidity.

Free acidity: It is only due to the presence of free HCl without stimulation.

Combined acidity: Presence of other acidic salts and organic acids like lactic acid, butyric acid etc gives combined acidity.

Total acidity: Free acidity + Combined acidity.

The aspirated gastric juice is titrated with the alkali like 0.1 N NaOH using the indicator like Topfer's indicator (0.5% dimethyl amino benzene in ethanol). The color changes from red to yellow. This indicates the free acidity level.

For total acidity measurement phenolphthalein indicator is used. The color changes from yellow to red again during titration. This indicates total acidity. The difference between total acidity and free acidity gives combined acidity.

2. Pentagastrin Stimulation Test

Pentagastrin is the synthetic peptide used for stimulation of gastric secretion. First the fasting gastric content is aspirated using Ryle's tube. This is known as Residual juice. Then the gastric juice is again collected after the next one hour. This represents the basal secretion. After that 5 µg per kg body of pentagastrin is given to the person to stimulate gastric secretion. The gastric juice is collected for one hour at each 15 minutes interval. This represents the maximum secretion. The collected samples are then analysed using phenolphthalein indicator for estimation of Basal acid output (BAO) and Maximal acid output (MAO).

In a conical flask, 5 ml of gastric juice is taken and a drop of phenolphthalein indicator is added. Then is titrated with 0.1N NaOH. Change of color to light pink indicates the endpoint of titration.

Basal acid output: Acid output under basal secretion. Normal value ranges from 1-2.5 mmol/hr.

Maximal acid output: Acid output after gastric stimulation using pentagastrin. Normal value ranges from 20-40 mmol/hr.

3. Augmented Histamine Test Meal

Histamin is a strong stimulator of gastric secretion. First the basal gastric juice is collected, then the stomach is stimulated by subcutaneous histamine administration (0.04 mg/kg body weight). The gastric contents are again aspirated and acidity in the samples is measured.

4. Tubeless Gastric Analysis

In this test, cation exchange resin is administered to the patient. It is quantitatively exchanged with the H^+ ions of gastric secretion. After the exchange, the resin is excreted in urine. This urine sample is collected and used for the measurement of gastric acidity indirectly.

Conditions for Increased Gastric HCl Secretion

- Zollinger-Ellison syndrome (tumor of gastrin secreting cells of pancreas).
- Duodenal ulcer.
- Histamine over production etc.

Conditions for Decreased Gastric HCl Secretion

- Gastric carcinoma.
- Gastritis.
- Pernicious anemia etc.

25

CSF ANALYSIS

CSF (Cerebro Spinal Fluid) is the fluid produced by the choroid plexus in the ventricle of brain. It is present in the cavity that surrounds brain and spinal cord between pia mater and arachnoid mater. The normal volume of CSF is 130 ml out of which 25 ml is present in ventricles.

Functions

- Medium to transfer substances from brain tissues and spinal cord to blood.
- Has Cushing effect (protects brain and spinal cord from injury)?

Clinical use: Diagnosis of central nervous system disorder that is reflected by meningitis (inflammation of meninges).

Physical Properties of CSF

1. Appearance: Clear, transparent, without any sediments or coagulum.

2. pH: 7.4-7.6.

3. Specific gravity: 1.006-1.007.

4. Pressure: 100-200 mm.

Chemical Properties of CSF

1. Protein: 15-45 mg/dl.

2. Sugar: 45-60 mg/dl.

3. Cell count: 0-4 × 10^6/L.

4. Urea: 20-40 mg/dl.

5. Calcium: 5.5-6 mg/dl.

6. Inorganic phosphate: 1.2-2.1 mg/dl.

7. Chloride: 120-130 mEQ/L.

8. IgG: 30 mg/L.

9. IgA: 4 mg/L.

Collection of CSF

It is collected by lumbar puncture between L3-L4 vertebrae under aseptic condition.

BIOCHEMICAL EXAMINATION OF CSF

Estimation of CSF Proteins

As the concentration of protein is very low in CSF, it is determined by turbidimetric method.

Principle

Sulphosalisylic acid causes protein precipitation resulting in turbidity. The turbidity of resultant uniform suspension is measured by colorimeter against standard at 660 nm.

Reagents

- 3% sulphosalisylic acid.
- Isotonic sodium chloride (8.8 gm/L).
- Standard protein (50 mg % of bovine serum albumin in 100 ml of isotonic sodium chloride).

Procedure

Reagent	Blank	Standard	Test
Standard protein	-	1 ml	-
CSF	-	-	1 ml
Isotonic NaCl	1 ml	-	-
3 % sulphosalicylic acid	5 ml	5 ml	5 ml

Mix and keep for 5 minutes. Then take OD at 660 nm.

Calculation

$$\text{Protein in 1 ml CSF} = \frac{\text{OD of test}}{\text{OD of standard}} \times \text{concentration of standard}$$

$$= \frac{\text{OD of test}}{\text{OD of standard}} \times 0.5$$

$$\text{Protein in 100 ml CSF} = \frac{\text{OD of test}}{\text{OD of standard}} \times 0.5 \times 100$$

=..........mg/dl

Estimation of Globulin in CSF (Pandy's Test):

Principle

Globulins can be precipitated using saturated solution of phenol in water. Water is absorbed to globulin molecules resulting in displacement of phenol from the solution. It causes fine and persistent turbidity.

Reagent: 10 gm of phenol in 150 ml distilled water

Procedure

Add 2 drops of CSF to 2 ml of reagent and note the turbidity. The test is reportedas no opalescence, opalescence, marked opalescence and turbidity.

Significance: Marked opalescence and turbidity is indicative of increased globulin. It is seen in multiple sclerosis and neurosyphilis.

Estimation of Glucose

It can estimate by various methods that are available for glucose estimation. Most reliable one is GODPOD method.

Estimation of CSF Chloride

It can be done by chloride meter which uses ion selective electrode.

CSF Comparision in Various Disorders

Condition	Appearance	Cell count	Protein	Sugar	Coagulat
Normal	Clear and colorless	0-4 × 10^6/L	15-45 mg/dl	45-60 mg/dl	No
Bacterial meningitis	Opalescent	↑↑	↑↑	↓↓	May clot standin
Tuberculous meningitis	May be opalescent	Lymphocytes and mononuclear cells	↑	↓	Presen (cobweb t
Viral infection	Clear and colorless	↑	↑	Normal	No
Brain tumor	Clear and colorless	Normal	↑	↓	Solidifi
Subarachnoid hemorrhage	Blood in fresh hemorrhage	RBCs and WBCs	↑	Not significant	No

26

CHROMATOGRAPHY

Introduction

- Chromatography is a technique used for separation of components of a mixture by the distribution between the two phases.
- Invented by Russian botanist Mithail Tswett.
- Two phases of media involves:
  - **Stationary phase:** Supporting medium on which mixture to be separated is adsorbed. e.g paper or silica gel.
  - **Mobile phase:** Mixture of solvents and carries components of mixture across the stationary phase.

Principle

Chromatography is based on two factors:

- Different rates of migration of components of a mixture in solvent system
- Different rates of adsorption of the components on the stationary phase
- Mobility of various components of the mixture differs in these two phases and the measure of this mobility is called Rf value (relative fraction value). It is calculated as:

$$Rf = \frac{\text{distance travelled by substance}}{\text{distance travelled by the solvent}}$$

Types of Chromatography:

A. Depending on the Mobile Phase

1. Liquid chromatography
   - i. Solid-liquid chromatography.
   - ii. Liquid-liquid chromatography.
2. Gas chromatography
   - i. Solid-gas chromatography.
   - ii. Liquid-gas chromatography.

B. Depending Upon the Actual Technique Used:

1. Planar chromatography

The vertical or horizontal plane in the form of paper, glass or plastic is used as supporting medium, which is saturated with solvent system as such like in paper chromatography (Whatman No. 1 filter paper as supporting medium) or first coated with thin layer of silica gel that then carries the stationary phase (e.g., TLC/thin layer chromatography).

2. Column chromatography:

Chromatographic medium is contained in a glass tube or a column.

C. Depending Upon The Mechanism Involved:

1. Adsorption chromatography

- ➢ Separation is based on the difference in the adsorption capacity of solutes for particular adsorbents e.g. silica, charcoal etc.

2. Partition chromatography

- ➢ Separation is based on difference in solubilities of solute between 2 liquid phases one of which is stationary phase and the other is mobile phase.
- ➢ Mainly used for diagnosis of aminoaciduria.

3. Affinity chromatography

- ➢ Separation is based on affinity of particular solute towards the stationary phase.
- ➢ Routinely used for the separation of antigens, HbA1c etc.

4. Gel filtration chromatography

- ➢ Separation is based on the size of solute particle.
- ➢ Solutes of high molecular weight will be excluded from the gel and that with low molecular weight will be trapped.
- ➢ Used to purify protein mixtures from salt, ions etc.

5. Ion exchange chromatography

- ➢ Separation is based on magnitude of charge on the particle.
- ➢ Used to separate organic and inorganic ions, proteins and nucleotides.

6. HPLC (high performance liquid chromatography)

Aim: To Study Paper Chromatography

Requirements

- Whatman no.1 filter paper and pencil.
- Test tubes with fitting cork or rubber stopper splitted in the middle.
- Amino acids alanine and methionine.
- Amino acid mixture containing alanine and methionine.
- Solvent mixture (n-butanol, acetic acid and distilled water in the ratio 12:3:5 respectively).
- Micropipette (1-5 μl).
- Ninhydrin solution 0.1%.

Procedure

- Wear gloves and cut Whatman no. 1 filter paper of appropriate size.
- Make a line with pencil 2.5 cm above the one end of the paper.
- Mark 3 small spots on the line with pencil. The distance between the spots should be at least 2 cm. And the spots should be at least 1.5 cm away from the edges of the paper.
- Take the solvent system in the chromatographic chamber upto the height of 2.5 cm and cover it with lid for 15-30 min so as to allow for the saturation of the chamber.
- Apply 5 μl of amino acid mixture in one of the spots marked with pencil and dry the spot with the drier. Similarly apply pure alanine and methionine solutions to the other spots separately.
- Put the paper carefully in the chromatographic chamber such that the sides of the paper do not touch the walls of the chamber and the sample spots do not come in contact with the solvent. The spots should be above the level of solvent.
- The paper should be kept in vertical position with the help of thread or adhesive tape or cork.
- Cover the chamber with lid and leave it for few hours. The solvent moves through the paper.
- When the solvent reaches near the upper end of the paper, remove the paper from the chamber and mark the solvent front with the pencil. Dry it in oven at 100-110 ^{0}C.
- Dip the paper in ninhydrin solution and dry it again.
- Purple colored spots appear in the paper.
- Calculate the Rf values.

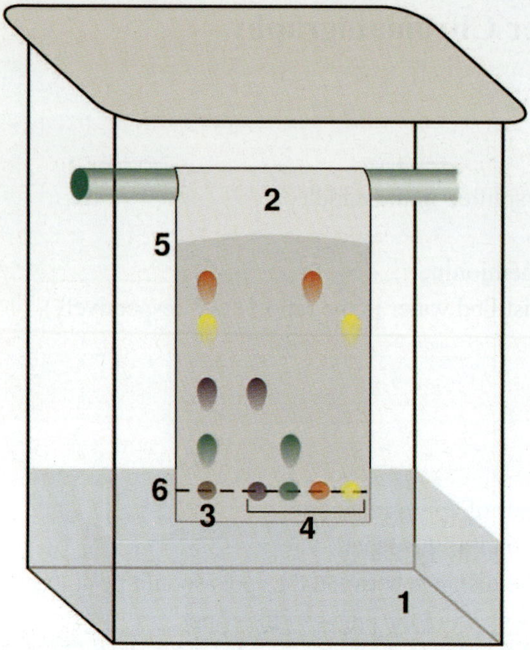

1= Buffer tank with buffer

2= Paper Strip

3= Spot of solution containing mixture of amino acids

4= Spots of solutions containing only one amino acid

5= Solvent front (distance travelled by solvent)

6= Origin or base line where sample is applied

Fig. 26.1: Chromatography

$$Rf = \frac{\text{Distance travelled by solute from origin}}{\text{Distance between } 6 - 5 \text{ (solvent front or distance travelled by solvent from origin)}}$$

27

ELECTROPHORESIS

INTRODUCTION

- The term electrophoresis refers to the migration of charged particles in a liquid medium under the influence of electric field.
- Arne Jiseluis introduced this technique.

Principle

The velocity of the migration of charged molecule in an electric field depends on:

- Net charge.
- Strength of electric field.
- Molecular weight of the particle.

Negatively charged ions move towards anode and positively charged ions move towards cathode. At pH 8.6, all the serum proteins carry net negative charge and hence move towards anode.

Support Media

1. **Inert media:** Paper (Whatman No.1 filter paper), cellulose acetate strip.

2. **Media as molecular sieve:** Agarose gel, polyacrylamide gel. These act as molecular sieve restricting the movement of larger particles.

Types of Electrophoresis

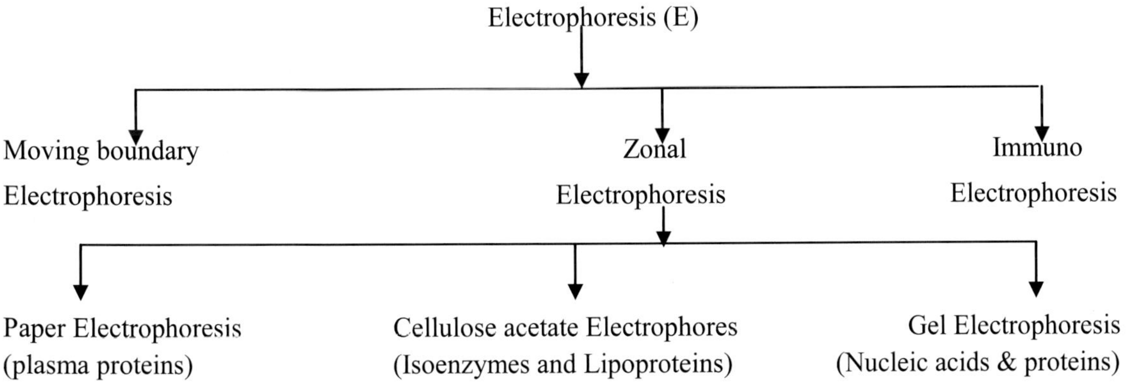

Factors Affecting Electrophoresis:

- Charge on ions.
- Size of charged particles.
- pH of the medium.
- Strength of the electric field.
- Properties of supporting medium.
- Temperature of medium.
- Stability of the particle.

Common stains used in electrophoresis

Stains	For
Amino black, Bromophenol blue	Plasma proteins
Sudan black B	Lipoproteins
Ethidium bromide	DNA fragments
Silver nitrate	Cerebro spinal fluid

Requirements

- Paper electrophoresis apparatus.
- Power supply device.
- Whatman No.1 filter paper.
- Barbitone buffer (pH 8.6).
- Bromophenol blue dye in alcohol.
- Dilute acetic acid.
- NaOH (0.05 N).

Procedure

- Cut the Whatman No. 1 filter paper into the strips of required size and mark a line with pencil about 5-6 cm from one end.
- Fill both the wells of clean and dry electrophoretic apparatus with buffer to equal levels.
- Wet the cut paper strip with buffer and remove excess buffer by pressing between the layers of dry filter paper.
- Put the buffer into the chamber and cover it with lid.
- Connect the chamber with power supply by connecting the cathode with the paper strip end with the pencil mark and anode to the other end.
- Switch on the power supply and adjust it by choosing either voltage 4.2 volt/cm or current 1.52 ampere.
- Allow to run for 10 minutes and then switch off the power supply.
- Apply 5 to 10 µl of serum sample using a micropipette on the strip at the pencil line.

- Cover the lid, switch on the power supply and allow it to run for 15–18 hours undisturbed.
- After completion switch off the power supply and remove the strip.
- Dry in oven at 100 °C for 10–15 minutes to denature and fix the proteins on paper.
- Stain the strip with bromophenol blue by dipping the strip in the dye for 10–15 minutes.
- Remove the strip and wash excess stain with dilute acetic acid 3-4 times.
- Air dry the strip and note the different bands of albumin, α1, α2, β and γ globulin fractions.

Densitometry

Once electrophoretic separation and staining are complete, direct densitometry is used to quantify the zones either as percentage or as absolute concentration if total quantity of protein is known.

Electrophoriogram

It displays protein zone. Each zone is separated from the neighboring zone on the electrophoretic support material.

Result

Albumin carries the largest charge and therefore moves farthest. γ globulin has least or smallest charge, hence moves the smallest distance.

Note: In many diseases such as acute or chronic infections (malaria, TB) and connective tissue disorders (rheumatoid arthritis, systemic lupus erythromatosus), there is generalized increase in immunoglobulins of various types produced by many clones of plasma cells. This gives large and intense staining but diffused band in γ globulin region. This is called polyclonal gammopathy. This is in contrast to sharp, intense and well defined band. M band is produced in multiple myeloma which is monoclonal gammopathy.

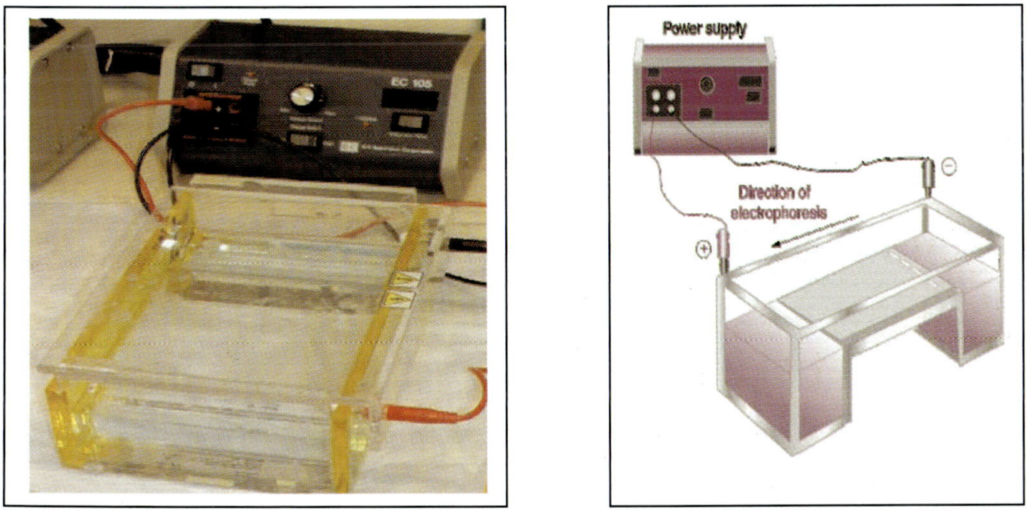

Fig. 27.1: Electrophoretic apparatus

Electrophoretic Patterns of Serum Protein on Various Conditions

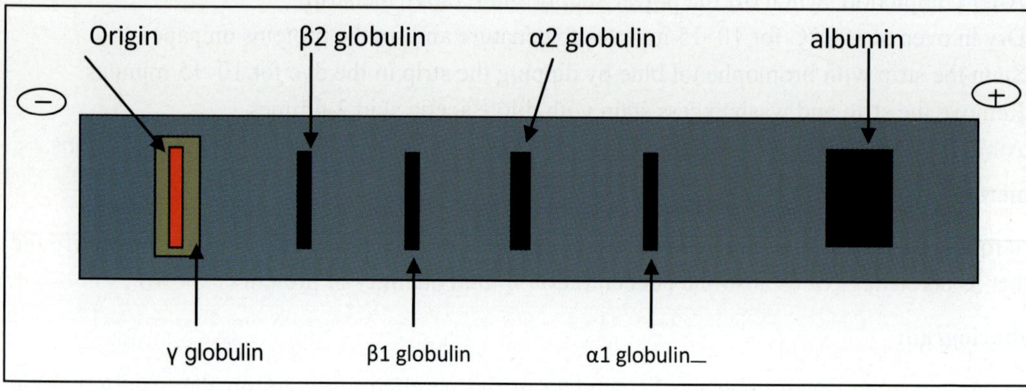

1. Normal

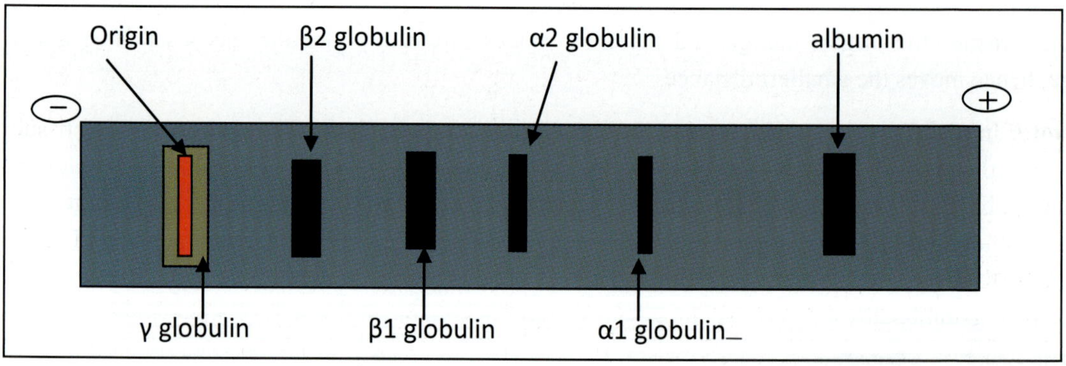

2. Liver cirrhosis

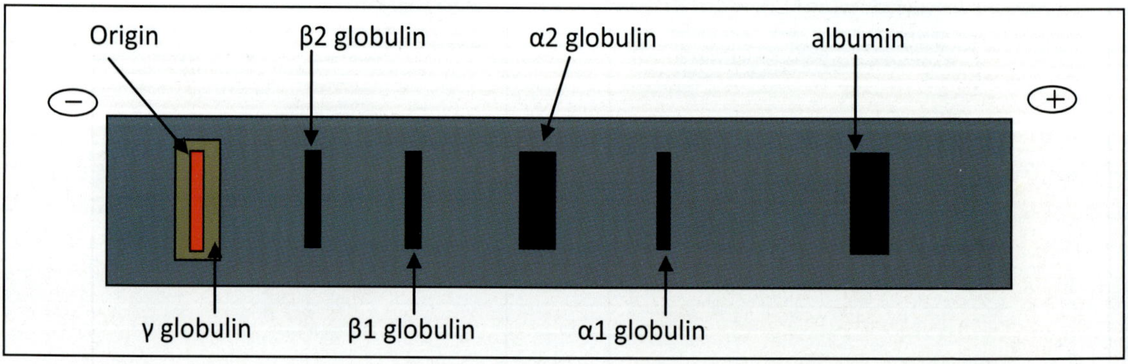

3. Nephrotic syndrome

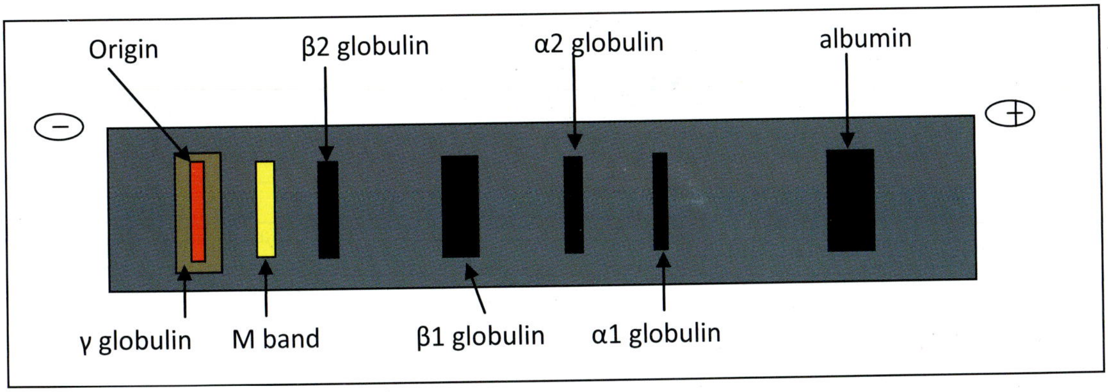

4. Multiple myeloma

28

FLAME PHOTOMETER

INTRODUCTION

- A flame photometer is a device used in inorganic chemical analysis to determine the concentration of certain elements that can be easily excited, e.g., sodium, potassium, lithium, calcium etc.
- It is based on the measurement of intensity of the light emitted when a metal is introduced into flame.
- The main use of flame photometer in clinical biochemistry is to determine level of sodium and potassium in serum.

Principle

When a liquid sample containing metallic salt (alkali and alkaline earth metal) solution is introduced into the flame:

- The solvent is vaporized leaving particles of the solid salt.
- The salt is vaporized or converted to gaseous state.
- A part or all of the gaseous molecules are progressively dissociated to give free neutral particles or radicals which are excited by the thermal energy of the flame.
- The excited atoms are unstable and return to ground state by emitting photons. Intensity of emitted radiation is proportional to the sample concentration.

Instrumentation

It consists of:

1. **Nebulizer:** It helps to spray sample into the burner.
2. **Flame:** It generates heat that volatilizes the element that becomes luminous when it returns to the ground state.
3. **Monochromator:** Simple filters that screen out all other wavelengths of light except the specific one emitted by the element being analyzed. Heat filters are also placed between the flame and monochromator filter to stop the passage of heat.
4. **Photodetector:** quantifies the emitted light by converting it to an electrical impulse which is then read by galvanometer.

Sample Preparation

Dilute 0.1 ml of serum with 9.9 ml of double distilled water.

Estimation of Sodium

- Switch on the instrument.
- Turn on the gas supply and ignite.
- Put on the air supply.
- Keep double distilled water in a beaker under the sample inlet and aspirate.
- Take the working standard and diluted serum in the separate beakers labeled accordingly.
- Keep blank in the sample inlet and set zero.
- Then read the test and standard.
- Note the reading.

Clinical Significance

Plasma sodium concentration: 136-145 meq/L.

Hypernatremia: Increase in plasma sodium concentration.

Causes of Hypernatremia

- Water depletion.
- Primary aldosteronism and in some cases of Cushing's syndrome (due to sodium retaining action of hormones).

Hyponatremia: Decrease in plasma sodium concentration.

Causes of Hyponatremia

- Vomiting, diarrhea, adrenal insufficiency, renal dysfunction, renal tubular acidosis, trauma, burns, acute and chronic diseases, intestinal obstruction, hyperglycemia and hyperlipidemia.

Estimation of Potassium

- Replace the Na filter by the K filter.
- By adopting the same procedure as for sodium, take the reading for the sample and the standard.

Interpretation

Plasma potassium concentration: 3.5-5.5 meq/L.

Hyperkalemia: Increase in plasma potassium concentration. It affects skeletal muscles and may cause profound myocardial infarction leading to cardiac arrest (8-10 meq/L).

Causes of Hyperkalemia

- Decreased renal excretion (renal failure, Addison's disease, urinary tract obstruction).
- Entry to extracellular space (hemolysis, burns).
- Metabolic acidosis.

Hypokalemia: Decreased plasma potassium level. It is associated with muscle weakness, increased neuromuscular excitability and decreased myocardial function.

Causes of Hypokalemia

- Increased renal excretion (hyperaldosteronism, Cushing's syndrome).
- Alkalosis
- Gastrointestinal loss (diarrhea, vomiting, malabsorption).

Fig. 28.1: Flame photometer

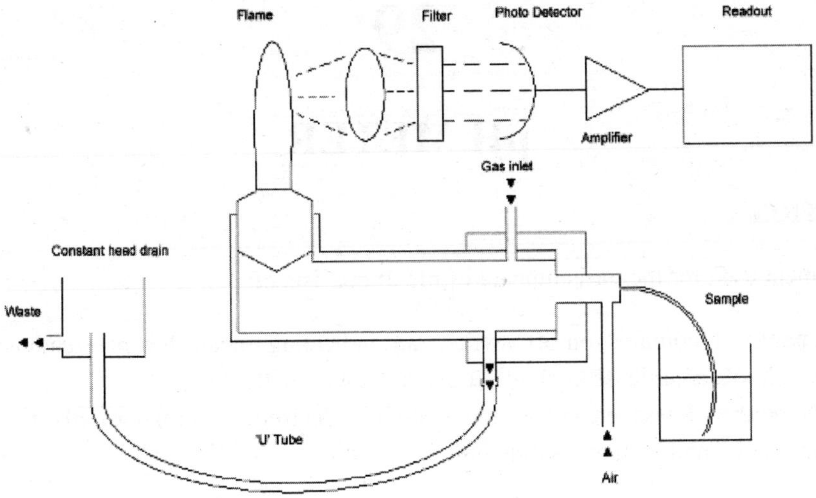

Fig. 28.2: Schematic diagram of flame photometer

29

pH METER

INTRODUCTION

It is the instrument used for the measurement of pH. It consists of:

1. **Power pack:** It contains on/off switch, an indicating meter for pH reading, temperature adjustment knob, a calibrate knob to set pH and a wire with plug pin.
2. **Two electrodes:** Glass electrode and Calomel electrode. In modern pH meter these two electrodes are combined into a single unit.

Glass Electrode

- Has a hard glass tube.
- At its base is a thin bulb that contains HCl (0.1 mol/l).
- Bulb is covered with special membrane of soda glass.
- Membrane is sensitive to H^+ ions and allows H^+ ions to pass through it.
- A platinum wire is connected to HCl through silver-silver chloride electrode.
- The wire coming out from the glass electrode is connected to the power pack of the pH meter.

Calomel Electrode

- It is a reference electrode which is not sensitive to H^+ ions.
- Consists of calomel paste that is connected with platinum wire through mercury.
- The free end of calomel electrode has a porous plug.
- The base of electrode consists of deposits of KCl crystals.
- The remaining portion is filled with KCl solution.

The glass electrode in a test solution constitutes a half cell and calomel electrode constitutes another half cell. The two electrodes complete the circuit.

Principle

pH meter is based on the measurement of electromotive force (e.m.f.) generated between the two electrodes due to difference in H^+ ion concentration.

Applications

- Measurement of pH of biological fluids.
- Petrochemical industry.
- Food and agriculture.
- pH of solution like sea water.
- Other chemical process.

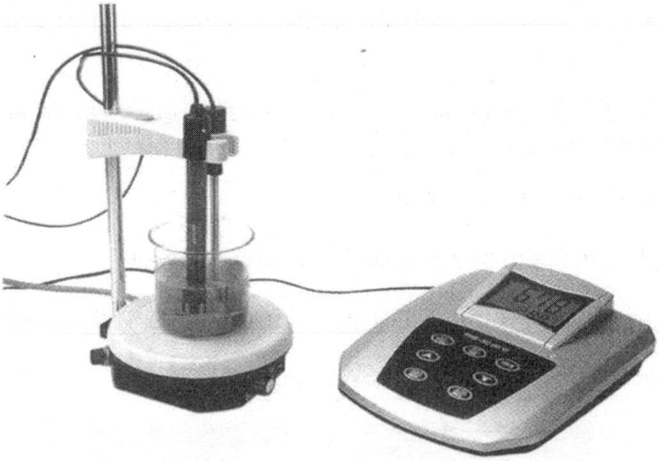

Fig. 29.1: pH meter

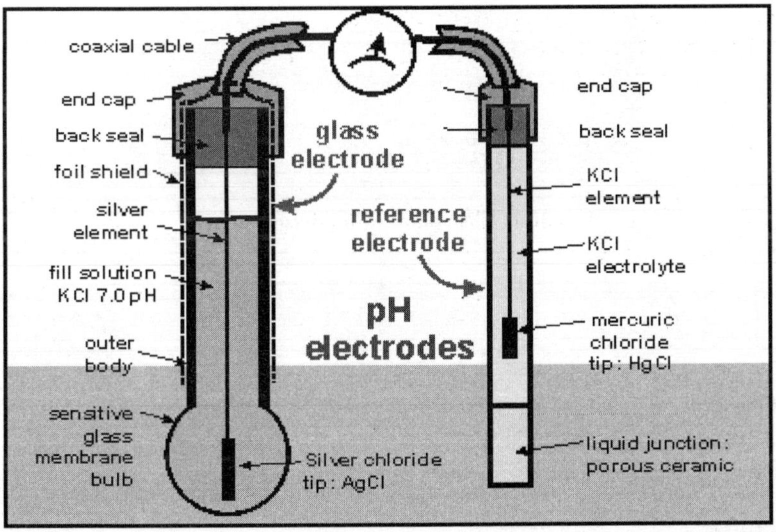

Fig. 29.2: Electrodes

30

ELISA

- It is known as Enzyme-linked immunoadsorbent assay.
- Commonly used to measure antibody, antigens, proteins and glycoproteins in biological samples.
- E.g., diagnosis of HIV, pregnancy tests, measurement of cytokines or soluble receptors in serum, food allergens.
- Generally carried out in 96 wells plate.
- Allows multiple samples to be measured in single experiment.
- Enzyme commonly used: Horse raddish peroxidase and ALP (alkaline phosphatase).
- Colored product formed is read by ELISA reader.

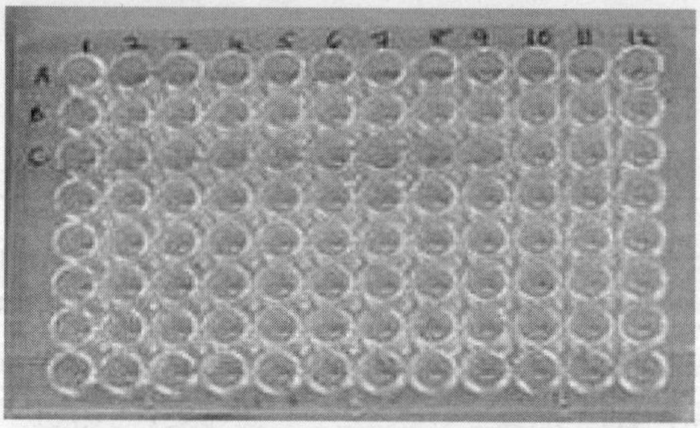

Fig. 30.1: Elisa plate with microtitre wells

Types

- Direct.
- Indirect.
- Sandwich.
- Competitive.

Direct ELISA (only a labelled primary antibody is used)

- Antigen is coated on surface of ELISA plate.
- Antibody (primary) with enzyme conjugate is added to the antigen.
- Antigen binds to antibody.
- Plate is washed to remove unbound antibody and substrate is added.
- Enzyme converts substrate to colored product.

Indirect ELISA

- Antigen is coated on surface of ELISA plate.
- Antibody (primary) is added to the antigen.
- Antigen binds to antigen.
- Microtiter plates are washed to remove unbound antibody.
- Secondary antibody (against primary antibody) conjugated with the enzyme is added.
- Bindings occur and plates are again washed to remove unbound fraction.
- Substrate for enzyme is added and colored product develops.

Sandwich ELISA (Antigen stuck between two antibodies)

- ELISA well surface is coated with known quantity of capture antibody.
- Sample containing antigen is added; binding occurs.
- Unbound antigen is removed by washing.
- Secondary antibody linked to enzyme is added.
- Plates are washed to remove unbound fraction.
- Then substrate to enzyme is added. Enzyme converts substrate to colored product.

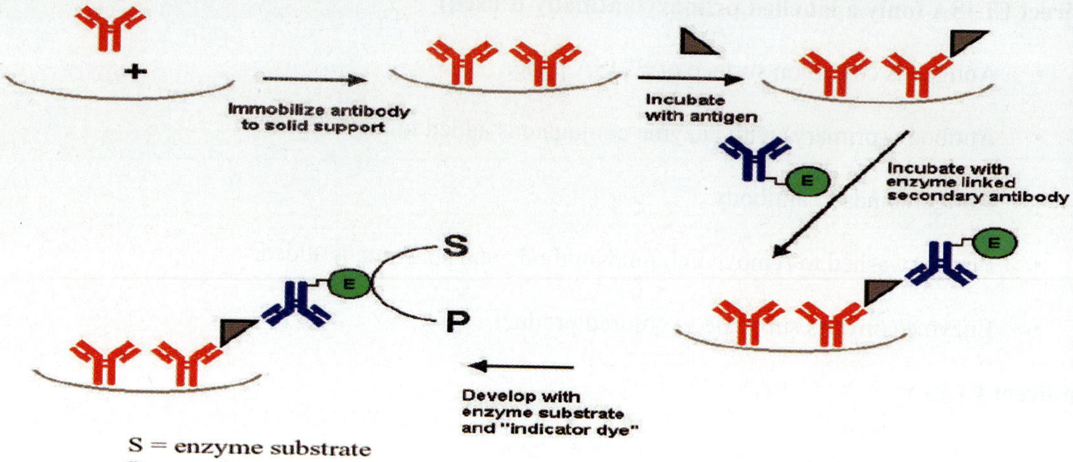

S = enzyme substrate
P = product

Competitive ELISA

- Unlabelled primary antibody first incubated with the antigen in sample.
- Ag-Ab complex then added to an antigen coated well of microtiter plate.
- Plate washed to remove unbound antibodies.
- More is the antigen present in the sample, more Ag-Ab complexes are formed so less free antibodies are available to bind antigen in the well and vice versa, hence there occurs competition.
- Secondary antibody (linked with enzymes) specific to primary antibody added followed by washing to remove unbound part.
- Substrate is added that is changed to colored product by enzyme.

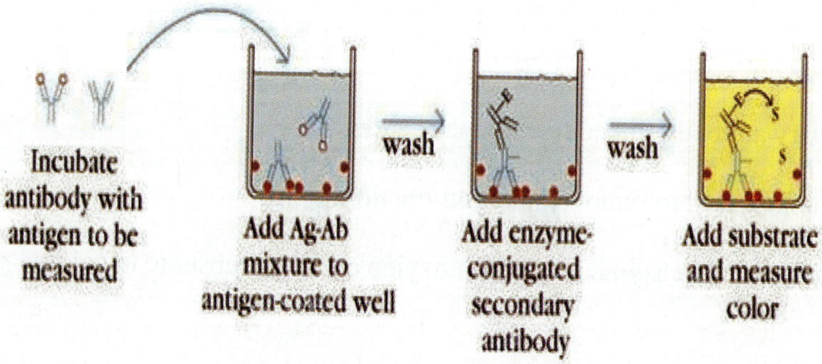

31

ION SELECTIVE ELECTRODE (ISE)

- Also known as indicator electrodes or specific ion electrodes that convert the activity of specific ion in sample to electrochemical potential.
- Respond directly to the analyte.
- Used for direct potentiometric measurements.
- Selectively binds and measures the activity of one ion.
- Responds preferentially to one species in solution.
- Made from selective ion-conducting membrane (ion-exchange material that allows ions of one electrical sign to pass through).
- Reference electrode is inbuilt.
- Internal solution (solution inside electrode) contains ion of interest with constant activity.
- Ion of interest is also mixed with membrane.
- Membrane is nonporous and water insoluble.

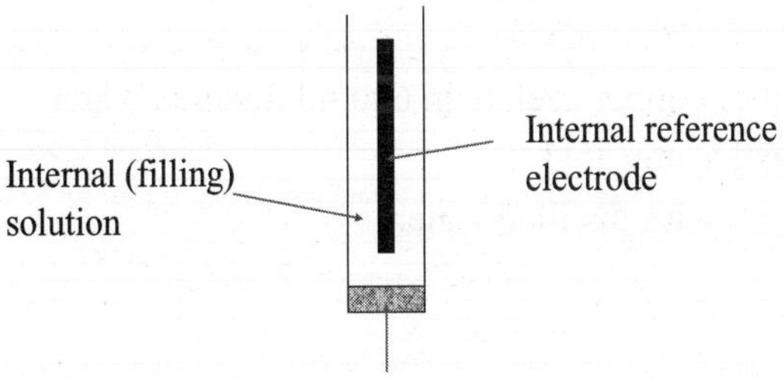

Fig. 31.1: Ion selective membrane

Appendix I

REAGENT PREPARATION

A. For Tests of Carbohydrates

1. α-Naphthol (1% in ethanol)

- Dissolve 1 gm of α-naphthol in 50 ml of ethanol.
- Then maintain volume to 100 ml by ethanol.

2. Iodine Solution (0.1N)

- **Stock solution:** Dissolve 1.27 gm of iodine and 3 gm of potassium iodide in 40-50 ml of distilled water and maintain the volume to 100 ml with distilled water.
- **Working solution:** Dilute stock by 1: 10 in distilled water (1ml of stock iodine solution + 9 ml of distilled water).

3. Benedict's Qualitative Reagent

- **Solution A:** Dissolve 100 gm of sodium citrate, 100 gm of anhydrous sodium carbonate in 400-500 ml of distilled water.
- **Solution B:** Dissolve 17.3 gm of copper sulphate pentahydrate ($CuSO_4$, $5H_2O$) in 100 ml of distilled water.
- Transfer the solution B to solution A with constant stirring.
- Maintain the volume to 1L with distilled water.

4. Barfoed's Reagent

- Dissolve 66.5 gm of copper acetate in 600 ml distilled water.
- Add 9 ml of glacial acetic acid.
- Make volume to 1L with distilled water.

5. Seliwanoff's Reagent

- Dissolve 50 mg of resorcinol in 33 ml of conc HCl.
- Dilute to 100ml with distilled water.

6. Phenylhydrazine Mixture (mixture of phenyl hydrazine hydrochloride and sodium acetate in 1:1 ratio).

B. For Precipitation and Color Reactions of Proteins

1. Copper Sulphate (1%)

- Dissolve 1 gm of copper sulphate in 50-60 ml of distilled water. Then maintain the volume to 100 ml with distilled water.

2. Sodium Hydroxide (5%)

- Dissolve 5 gm of sodium hydroxide in 50-60 ml of distilled water and then maintain the volume to 100 ml with distilled water.

3. Ninhydrin Solution (0.2%)

- Dissolve 0.2 gm of ninhydrin in 40-50 ml of distilled water. Then maintain the volume to 100 ml with distilled water.

4. Sodium Hydroxide (40%)

- Dissolve 40 gm of sodium hydroxide in 50-60 ml of distilled water. Then maintain the volume to 100 ml with distilled water.

5. Millon's Reagent

- Dissolve 10 gm of mercuric sulphate in 40-50 ml of distilled water.
- Add 10 ml of conc H_2SO_4.
- Make volume to 100 ml with distilled water.

6. Sodium Nitrite (1%)

- Dissolve 1 gm of sodium nitrite ($NaNO_2$) in 40-50 ml distilled water and maintain the volume to 100 ml with distilled water.

7. Formalin (0.2%)

- Add 1 ml of formalin in 499 ml of distilled water (1:500).

8. Mercuric Sulphate (10%)

- Dissolve 10 gm of mercuric sulphate in 40-50 ml distilled water and maintain the volume to 100 ml with distilled water.

9. Sodium Hypobromite (10%)

- Add 10 ml of bromine in 90 ml of 10% NaOH (10 gm of NaOH maintained to 100 ml with distilled water).

10. Lead Acetate (10%)

- Dissolve 10 gm of lead acetate in 40-50 ml of distilled water. Then maintain the volume to 100 ml with distilled water.

11. Sulphanilic Acid (0.5% in 2% HCl)

- 2% HCl: Add 6 ml conc HCl in 100 ml of distilled water.
- Dissolve 500 mg of sulphanilic acid in 40-50 ml of 2% HCl and maintain the volume to 100 ml with 2% HCl.

12. Sodium Carbonate (10%)

- Dissolve 10 gm of sodium carbonate in 40-50 ml of distilled water. Then maintain the volume to 100 ml with distilled water.

13. Isoelectric Precipitation (Caesin)

Sodium Acetate (0.1 M)

- Dissolve 8.2 gm of anhydrous sodium acetate or 13.6 gm of sodium acetate trihydrate in 50-60 ml of distilled water. Then maintain the volume to 100 ml.

Casein Solution (1%)

- Dissolve 1gm of casein in 100 ml of 0.1M sodium acetate.

Acetic Acid (1N)

- Take 57.5 ml of glacial acetic acid and dilute it to 1L with distilled water.

Acetic Acid (0.1)

- Dilute 1 N acetic acid ten times (10 ml of 1N acetic acid diluted to 100 ml with distilled water).

For Urine Analysis

1. Benedict's Qualitative Reagent

- **Solution A:** Dissolve 100 gm of sodium citrate, 100 gm of anhydrous sodium carbonate in 400-500 ml of distilled water.
- **Solution B:** Dissolve 17.3 gm of copper sulphate pentahydrate ($CuSO_4$, $5H_2O$) in 100 ml of distilled water.
- Transfer the solution B to solution A with constant stirring.
- Maintain the volume to 1L with distilled water.

2. 20% Sulphosalicylic Acid

- Dissolve 20 gm sulphosalicylic acid in 50-60 ml of distilled water and maintain the volume to 100 with distilled water.

3. 5% Sodium Nitroprusside

- Dissolve 5 gm of sodium nitroprusside in 50-60 ml distilled water and then maintain the volume to 100 ml with distilled water.

4. Bile Salt (0.1%)

- Dissolve 0.1 gm of sodium glycocholate or sodium taurocholate in 60 ml of distilled water then make the volume to 100 ml with distilled water.

5. Fouchet's Reagent

- **Solution 1:** Dissolve 250 gm of trichloroacetic acid in 500 ml of distilled water.
- **Solution 2:** Dissolve 10 gm of $FeCl_3$ in 100 ml of distilled water.
- Mix solutions 1 and 2. Then make the volume to 1L with distilled water.

6. Erhlich's Reagent

- Dissolve 250 gm of dimethylaminobenzaldehyde in 150 ml conc HCl. Then maintain the volume to 1L with distilled water.

7. Benzidine (saturated solution in glacial acetic acid).

8. Sodium Acetate (saturated solution in distilled water).

For Colorimetry

1. Stock $KMnO_4$ (100 mg/dl)

Maintain 100 mg of $KMnO_4$ to 100 ml with distilled water.

BLOOD SUGAR ESTIMATION

For King and Asatoor Method

1. Alkaline Copper Sulphate Solution

Solution 1:

- Dissolve 13 gm of $CuSO_4$ in 500 ml of distilled water and then maintain the volume to 1L.

Solution 2:

- Dissolve 40 gm of sodium carbonate and 50 gm of sodium bicarbonate in 300 ml of distilled water.
- Separately dissolve 24 gm of sodium potassium tartarate and 36.8 gm of potassium oxalate in 300 ml of distilled water.
- Mix these two solutions and maintain the final volume to 1L with distilled water.

Mix equal volume of solution 1 and 2 (mix freshly before use)

2. Isotonic Copper Sulphate

- Dissolve 3.8 gm of sodium sulphate and 7 gm copper sulphate in 50 ml of distilled water and make up the volume to 100 ml with distilled water.

3. Phosphomolybdic Acid

- Dissolve 40 gm of NaOH, 70 gm of molybdic acid and 10 gm of sodium tungstate in 300 ml distilled water and boil for 1 hour.
- Cool this solution and add 250 ml of orthophosphoric acid. Then maintain the volume to 1L with distilled water.

4. Sodium Tungstate (10%)

- Maintain 10 gm of sodium tungstate to 100 ml with distilled water.

5. Standard Glucose (100 mg/dl)

- Dissolve 100 mg of glucose in 50 ml of distilled water and maintain volume to 100 ml with distilled water.

For O-Toluidine Method

1. O-Toluidine Reagent

- Dissolve 5 gm of thiourea in 90 ml of O-toluidine and make the volume to 1L with glacial acetic acid.

2. Trichloroacetic Acid (10%)

- Maintain 10 gm of trichloroacetic acid to 100 ml with distilled water.

3. H_2SO_4 (2/3N)

- Dilute 18.3 ml conc H_2SO_4 to 1L with distilled water.

4. Standard Glucose (100 mg/dl)

- Dissolve 100 mg of glucose in 50 ml of distilled water and maintain volume to 100 ml with distilled water.

Serum Creatinine Estimation (Jaffe's Method)

1. Sodium Hydroxide (5%)

- Dissolve 5 gm of sodium hydroxide in 50-60 ml of distilled water and then maintain the volume to 100 ml with distilled water.

2. H_2SO_4 (2/3N)

- Dilute 18.3 ml conc H_2SO_4 to 1L with distilled water.

3. Stock Creatinine (100 mg/dl)

- Dissolve 100 mg of creatinine in 50 ml of distilled water and maintain volume to 100 ml with distilled water.

4. Standard Creatinine (2 mg/dl)

- Dilute 2 ml of stock to 100 ml with distilled water.

5. Saturated Picric Acid

For Urea Estimation

1. Acid Reagent

Reagent 1:

- Dissolve 5 gm of $FeCl_3$ in 50 ml of distilled water. Then slowly add 100 ml of orthophosphoric acid and maintain final volume to 250 ml with distilled water.

Reagent 2:

- Take 800 ml of distilled water and slowly add 200 ml of conc H_2SO_4 to it.

Add 1ml of Reagent 1 to 999 ml of Reagent 2. Mix and keep in refrigerated condition in dark bottle.

2. Coloring Reagent

i. Diacetylmonoxime (2.5%)

- Dissolve 2.5 gm of diacetyl monoxime (DAM) in 40 ml dstilled water and maintain the volume to 100 ml.

ii. Thiosemicarbazide (0.5%)

- Dissolve 0.5 gm of thiosemicarbazide in 40 ml distilled water and maintain the volume to 100 ml.

Mix 67 ml of 2.5% DAM and 67 ml of 0.5% thiosemicarbazide. Then maintain the volume to 1L with distilled water

3. Stock Urea Solution (100 mg/dl)

- Maintain 100 mg of urea to 100 ml with distilled water.

4. Urea Standard (1 mg/dl)

- Dilute 1 ml of stock to 100 ml with distilled water.

For Uric Acid Estimation

1. Sodium Tungstate (10%)

- Maintain 10 gm of sodium tungstate to 100 ml with distilled water.

2. H_2SO_4 (2/3N)

- Dilute18.3 ml conc H_2SO_4 to 1L with distilled water.

3. Tungstic Acid Reagent

- Mix 50 ml of 10% sodium tungstate to 50 ml of 2/3N H_2SO_4. Then maintain final volume to 800 ml with distilled water. Add few drops of phosphoric acid and preserve.

4. Phosphotungstic Acid Reagent

Stock Solution:

- Dissolve 50 gm of sodium tungstate in 300 ml of distilled water and add 40 ml of 85% phosphoric acid. Reflux for 2 hours and then make final volume to 500 ml with distilled water.

Working Solution

- Dilute 10 ml of stock to 100 ml with distilled water.

5. Uric Acid Standard

Stock uric acid (100 mg/dl)

- Maintain 100 mg of uric acid to 100 ml with distilled water.

Standard (1mg/dl)

- Dilute 1ml of stock uric acid solution to 100 ml with distilled water.

FOR PROTEIN ESTIMATION (Total Protein, Albumin and A:G Ratio)

Biuret Method for Total Protein

1. 0.2N NaOH (8 gm/L)

- Dissolve 8 gm of NaOH in 1L of distilled water.

2. Biuret Reagent for Total Protein Estimation

- **Stock:** Dissolve 45 gm of sodium potassium tartarate in 400 ml of 0.2N NaOH. After that add 15 gm of $CuSO_4$ and 5 gm of potassium iodide. Then make the final volume to 1L with 0.2N NaOH.
- **Working:** Dissolve 5 gm of potassium iodide in 200 ml of stock biuret reagent and then dilute to 1L with 0.2N NaOH.

3. Protein Standard (0.6 gm/dl)

- Dissolve 600 mg of bovine serum albumin in 50 ml of distilled water and maintain the final volume to 100 with distilled water.

BCG Method for Albumin

1. 1N NaOH

- Dissolve 40 gm of NaOH in 300 ml distilled water and maintain the final volume to 1L with distilled water.

2. BCG Reagent

i. *Succinate buffer:* Dissolve 56 gm of succinic acid and 10 gm of NaOH in 800 ml distilled water. Then adjust the pH to 4.1 with 1N NaOH and make final volume to 1L with distilled water.
ii. *BCG solution:* Dissolve 700 mg of BCG dye in 80 ml of distilled water and add 2 ml of 1N NaOH. Then make final volume to 100 ml with distilled water.
iii. *Sodium azide (4 gm/dl):* Maintain 4 gm of sodium azide to 100 ml with distilled water.
iv. *Brij 35 (25 gm %):* Maintain 25 gm of brij 35 to 100 ml with distilled water.
v. *BCG dye reagent:* Add 100 ml of succinate buffer (i), 8 ml of BCG solution (ii), 2.5 ml of 4 gm% sodium azide and 2.5 ml of 25 gm% brij 35 solution. Then make final volume to 1L with distilled water. The pH of solution so prepared should be 4.1 at 20 °C.

3. Protein Standard (0.6 gm/dl)

- Dissolve 600 mg of bovine serum albumin in 50 ml of distilled water and maintain the final volume to 100 with distilled water.

For Cholesterol Estimation

1. Acetic Acid-Ferric Chloride Reagent

- Dissolve 50 mg of anhydrous ferric chloride in 100 ml of glacial acetic acid.

2. Standard Cholesterol (200 mg/dl)

- Dissolve 200 mg of cholesterol in 50 ml of glacial acetic acid and maintain the volume to 100 ml with glacial acetic acid.

3. $FeCl_3$ Color Reagent

Stock: Dissolve 2.5 gm of $FeCl_3$ in 50 ml of orthophosphoric acid and make the volume to 100 with orthophosphoric acid.

Working: Add 8 ml of stock to 92 ml of H_2SO_4.

For Bilirubin Estimation

1. Diazo Blank

- Dilute 15 ml of concentrated acid HCl to 1L with distilled water.

2. Diazo A

- Dissolve 1 gm of sulphanilic acid in 400 ml of distilled water and add 15 ml of concentrated HCl. Then maintain the final volume to 1L with distilled water.

3. Diazo B

- Dissolve 0.5 gm of sodium nitrite in 50 ml of distilled water. Then maintain the final volume to 100 ml with distilled water. Prepare freshly before use.

4. Diazo Reagent

- Mix 10 ml of Diazo A and 0.3 ml of Diazo B.

5. Bilirubin Standard (10 mg/dl)

- Maintain 10 mg of bilirubin to 100 ml with distilled water. Store in dark bottle. Keep away from light as it is photosensitive and unstable.
- Alternatively methyl red can be used if bilirubin is not available.
- 290 mg methyl red in 100 ml distilled water=8 mg/dl of bilirubin standard with OD 0.32.

Appendix II

CASE STUDY

1. A person on routine checkup had following lab findings.

Fasting sugar	70 mg/dl	* (urine)
Post prandial sugar	138 mg/dl	** (urine)

There were no symptoms of polyuria, polydypsia and polyphasia.

i. What is probable diagnosis.
ii. Renal threshold for glucose.

2. A patient with acute abdominal pain had biochemical investigations such as Serum amylase, Serum lipase and Urinary amylase. All the values were highly increased.

i. What is the probable diagnosis?
ii. Give normal level of serum amylase and lipase.

3. A 10 years old child showed symptom like painful, swollen and bleeding gums; Petechiae; swollen and painful joints.

i. What may be the diagnosis?
ii. The disorder is caused due to deficiency of.
iii. Name the rich source of this deficient compound.

4. One of the college students was on hunger strike for few days. He was admitted to hospital in unconscious state with fruity odor. His laboratory findings were:

Blood sugar : 50 mg/dl	Blood pH: 7.22
Serum HCO_3^- :15 mEg/L	Urinary ketone bodies: **

i. What is the probable diagnosis?
ii. What is normal blood pH?
iii. Give the normal blood sugar level.
iv. Give the name of ketone bodies.
v. What is the cause of fruity odor?

5. A 30 years old female had following symptoms:

- ➢ High pigmentation around neck.
- ➢ Bright reddish patches on feet, ankle and face. These patches increased on exposure to sunlight.

i. What is the name of disorder?
ii. Name the cause of disorder.
iii. Mention the coenzyme forms of this compound.

6. A person was admitted to hospital in coma state. His lab findings were:

Blood sugar: 290 mg/dl | Sugar in urine: **

Serum bicarbonate: 15 mEq/l | Ketones in urine: ***

pH of blood: 7.23

i. What is the probable diagnosis?
ii. Name the test to detect ketone bodies.
iii. Give examples of ketone bodies.
iv. What are normal bicarbonate level and blood pH?

7. A person with a history of alcohol abuse was admitted to hospital after vomiting blood (hematemesis) on exercise. On examination, his liver was felt firm and enlarged. Serum levels of ALT, AST and bilirubin were high but level of serum total protein was low.

i. Write the probable diagnosis.
ii. What is the cause of hematemesis?
iii. What is the cause of increased transaminases and bilirubin?

8. Following were the result of OGTT of a patient.

Time	Blood glucose	Urine
Fasting	70	Nil
30 min	125	*
1 hr	155	***
1:30 hr	90	Nil
2 hr	80	Nil

i. Write the diagnosis.
ii. What is the renal threshold for glucose?
iii. Normal FBS =
iv. Normal PPBS=

9. An 11 years old girl had the history of poor vision in dim light and dryness of skin. When examined, she was found to be malnourished with grayish white spots on conjunctiva.

i. Mention the probable diagnosis.
ii. Mention the cause of disease.
iii. RDA of the compound deficient.

10. A newborn boy child developed yellowish skin and conjunctiva 2 days after birth. He had following lab reports:

Total serum bilirubin=19 mg /dl

Unconjugated bilirubin=15 mg/dl

- What is probable diagnosis?
- What may be done for treatment?
- Normal serum bilirubin level.

11. Following three conditions are given with their lab findings:

	A	B	C
Serum bilirubin	11 mg %	10 mg %	10 mg %
Conjugated bilirubin	0.6 mg %	5 mg %	8 mg %
Unconjugated bilirubin	10.5 mg %	5 mg %	2 mg %
ALP	9 KA	18 KA	150 KA
AST	26 units	270 units	70 U/L
ALT	23 units	295 units	80 U/L
Urine bile salt	-	**	**
Bile pigment in urine	-	*	**
Urobilinogen	***	*	-
Stercobilinogen	***	*	-

i. What is the probable diagnosis of these three cases?
ii. Give the names of bile salt and bile pigments.
iii. Normal level of ALT, AST, ALP.

12. A nine-month-old child urinated in a diper. When his mother was changing the diper after some time, she found that the diper stained black. On further investigations, the findings obtained were:

*Positive Benedict's test

**Negative GOD POD Test in urine

***Positive $FeCl_3$ Test

i. Write name of disorder.
ii. Give the enzyme deficient.
iii. Why diper stained black?
iv. Give the cause for positive Benedict's test.

13. A 45 years old patient had following findings:

Puffiness in face
Serum creatinine: 6 mg/dl
Serum urea: 90 mg/dl
Cholesterol: 400 mg/dl

Total protein: 4.2 g/dl
Albumin: 1.1 g/dl
Globuline: 3 g/dl
Urine protein: 9 g/L

i. What is probable diagnosis?
ii. Give normal A: G ratio.
iii. Give the normal cholesterol level.
iv. What is the cause of proteinuria?

14. The 12 years old child had symptoms like bow legs, rickety rosary, knee knocking during walk with following lab findings:

Serum calcium: 7 mg % Phosphorus 2.2 mg% ALP= 85 KA

i. What is probable diagnosis?
ii. What is compound is deficient?
iii. Active form of this compound is known as.
iv. What is normal level of calcium and phosphorus?

15. Interpret the result.

Serum calcium: 3.1 mmol/L (normal=2.1-2.6 mmol/L)

Serum phosphorous: 0.89 mmol/L (normal=1-1.5 mmol/L)

ALP: 18 KA units (normal=3-13 KAU)

i. What might be the disorder?
ii. What is the cause?

16. A female child of 3 years was mentally retarded with delayed milestones. On further examination the blood phenyl alanine level was very high and the urine of child imparted mousy odor.

i. What disease is the child suffering from?
ii. What is the cause?
iii. Why the urine imparted mousy odor?
iv. Name the screening test for the disease.

17. A 10 year old boy had a complain of difficulty to see in evening and night. He had diarrhea with recurrent infections. He looked malnourished with dry and rough skin. He used to eat mainly cereals, pulses but less fruits and green vegetables.

i. What is the diagnosis?
ii. What is the cause of this disorder?
iii. Mention the sources.

18. A person was admitted to hospital after acute chest pain. The biochemical findings were

Blood sugar: 270 mg/dl

Cholesterol: 350 mg/dl

HDL: 17 mg/dl

SGOT: 60 IU/L

SGPT: 25 IU/L

Creatine kinase and LDH elevated.

i. What is the diagnosis?
ii. Normal level of cholesterol and HDL.

19. A 17 years female had lethargic behavior. She had the complaints of weakness and fatigue with heavy menstrual bleeding. Her heart, lungs and abdomen were normal but she gave a pale appearance. On examination her lab report showed:

Hb: 8 gm%

TLC: 6700 cu/mm

RBC: 4.1×10^{12}/L

Blood smear: RBC normocytic and hypochromic

Serum iron: 46 ng/dl (50-150)

i. What is the probable diagnosis?
ii. What is normal Hb level?
iii. What is normal iron level?

20. A 45 years old female was obese with complaints of lethargy, cold intolerance, anemia and bradycardia. She went for thyroid checkup and her reports showed:

TSH: 25 mIU/L

fT4: 3 pmol/L

fT3: 4 pmol/L

i. What might be the case?
ii. What is normal thyroid hormone level?
iii. Why there is weight gain and cold intolerance?
iv. What is the cause for anemia?

Appendix III

ANSWERS TO CASES

1. i. Renal glycosuria

   ii. 180 mg/dl

2. i. Acute pancreatitis

   ii. Amylase: 50-120 IU/L

   Lipase: Up to 190 U/L

3. i. Scurvy

   ii. Vitamin C

   iii. Gooseberry, lemon, citrus fruits

4. i. Ketoacidosis

   ii. Blood pH=7.35-7.45

   iii. Fasting= 60-100 mg/dl

   Post prandial= 100-140 mg/dl

   Random= 60-140 mg/dl

   iv. Due to acetone (ketone body)

5. i. Pellagra

   ii. Deficiency of Niacin or Vitamin B3

   iii. NAD, NADP

6. i. Diabetic ketoacidosis

   ii. Rothera's test

   iii. Acetone, Acetoacetate, Beta hydroxybutyrate

iv. Blood pH= 7.35-7.45

HCO_3^-=21-28 mEq/L or 21-28 mmol/L

7. i. Liver cirrhosis

ii. Heavy physical exercise cause rupture of dilated and weak veins of esophagus. Damaged liver cannot produce sufficient amount of clotting factors causing regurgitation of blood. This leads to hematemesis.

iii. ALT and AST are released on cellular damage resulting in elevated levels. Bilirubin is conjugated in liver and excreted in bile. Damage to liver affects bilirubin metabolism causing increase in level.

8. i. Renal glycosuria

ii. 180 mg/dl

iii. FBS = 60-100 mg/dl

iv. PPBS= 100-140 mg/dl

9. i. Xerophthalmia

ii. Deficiency of vitamin A

iii. RDA= 5000 IU

10. i. Neonatal jaundice/physiological jaundice

ii. Phototherapy because in presence of light bilirubin is converted to non-toxic lumirubin that is readily excreted. This process is known as photoisomerization.

iii. Total bilirubin= 0.2-1 mg/dl

Unconjugated/Indirect bilirubin= 0.1-0.6 mg/dl

Conjugated/Direct bilirubin= 0.1-0.4 mg/dl

11. i. A= Hemolytic jaundice B= Hepatic Jaundice C= Obstructive jaundice

ii. Bile salt= Sodium glycochoate or sodium taurocholate Bile pigment= Bilirubin, Biliverdin

iii. ALT= 5-40 IU/L AST= 5-40 IU/L ALP= 40-125 IU/L

12. i. Alkaptonuria

ii. Homogentisate oxidase (Homogentisic acid cannot beconverted to Malely acetoaceate)

iii. Homogentisic acid is oxidized by polyphenol oxidase to benzoquinone acetate which is polymerized to black colored alkaptone bodies.

iv. Due to presence of homogentisic acid in urine.

13. i. Nephrotic syndrome.

ii. 1.5:1-2.5:1

iii. 150-200 mg/dl

iv. In nephrotic syndrome, kidneys have pores large enough to allow passage of proteins (but not cells)

14. i. Rickets

ii. Vitamin D

iii. Calcitriol

iv. Ca= 9-11 mg/dl P= 3-4 mg/dl

15. i. Hyperparathyroidism.

ii. Parathyroid hormone (PTH) is hypercalcemic hormone. It increases the level of plasma calcium and alkaline phosphatase (ALP) whereas decreases the level of inorganic phosphate. It causes bone resorption, increases renal tubular reabsorption of calcium but prevents the reabsorption of phosphate.

16. i. Phenylketonuria.

ii. Deficiency of enzyme phenylalanine hydroxylase.

iii. Due to deficiency of enzyme, phenylalanine cannot be converted to tyrosine. Phenylalanine is converted to alternate metabolites like phenylpyruvate, phenyl lactate and phenyl acetate which are then excreted in urine. Phenyl lactate imparts mousy odor to urine.

iv. Guthrie test.

17. i. Night blindness.

 ii. Deficiency of vitamin A.

 iii. Fish oil, milk, butter, green leafy vegetables, papaya, carrot, mango.

18. i. Myocardial infarction

 ii. Cholesterol=150-200 mg/dl HDL= 35-65 mg/dl

19. i. Iron deficiency anemia

 ii. Hb= 14-16 gm% (male) and 11-14 gm% (female)

 iii. Iron= 50-160 μg/dl or 9- 28.6 μmol/L

20. i. Primary hypothyroidism

 ii. fT3= 3.5-10 pmol/lL fT4=10.2-29.4 pmol/L

 TSH= 0.5-5 μU/ml or 0.5 -5 mIU/L

iii. Thyroid hormones have calorigenic effect. Deficiency would cause decreased cellular metabolism like oxidation of fat, carbohydrates and less heat production. This results in cold intolerance and obesity.

Appendix IV

SPOTTERS

1.

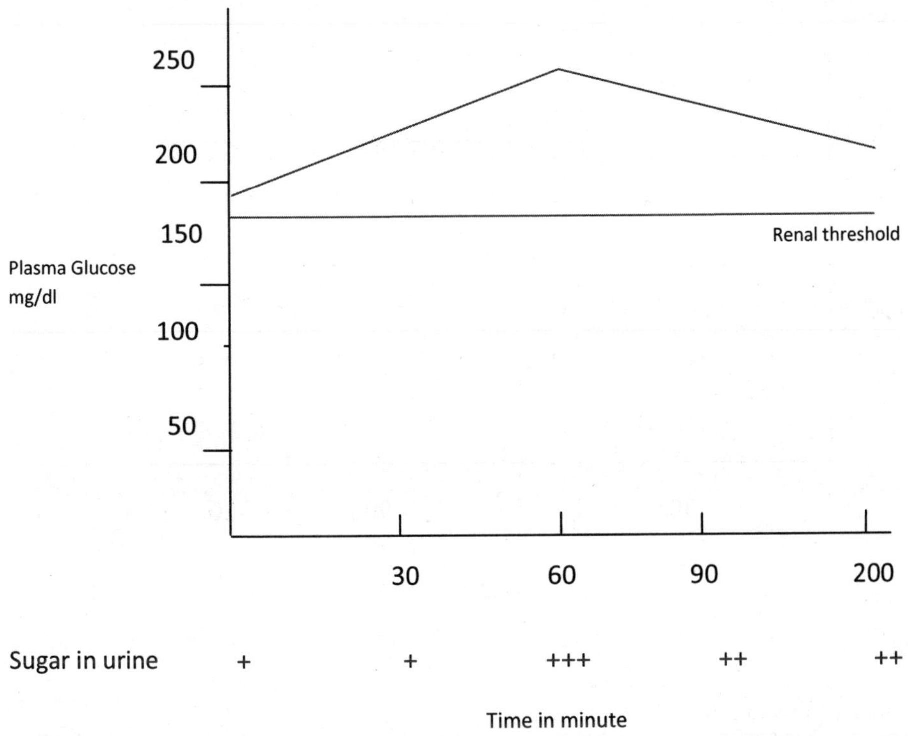

i. Identify the graph.

ii. Give the significance of this graph.

Answer

i. Oral glucose tolerance test curve indicating diabetes mellitus.

ii. Fasting plasma glucose (FBG)>126 mg/dl and 2 hours post load glucose >200 mg/dl. Blood glucose level rises beyond the renal threshold in diabetes that does not return to normal even after two hours of meal.

2.

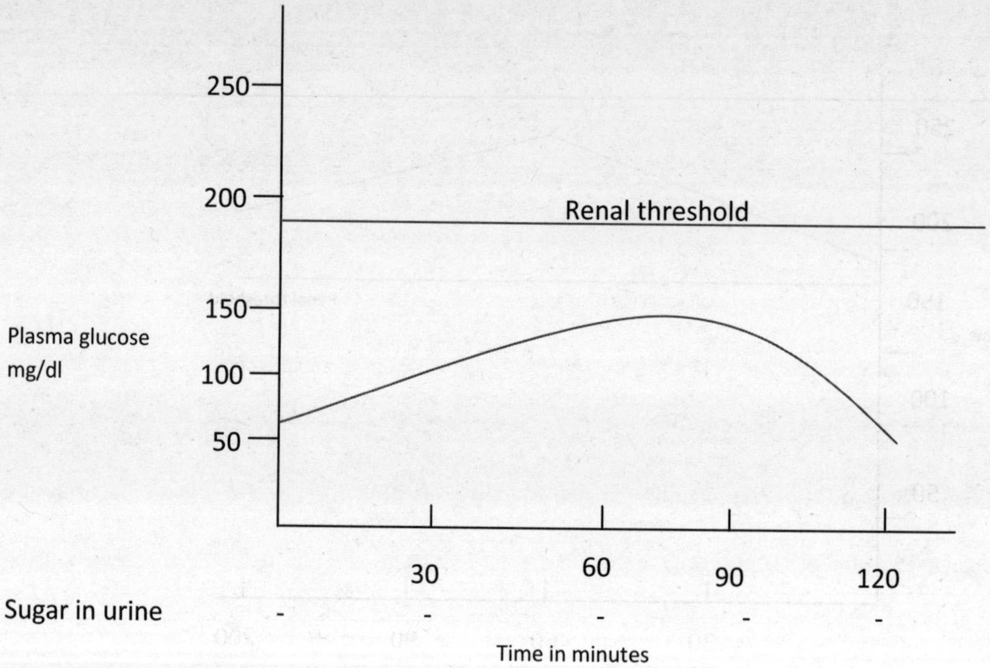

i. Identify the graph.

ii. Give the significance of this graph.

Answer

i. Normal oral glucose tolerance test curve.

ii. Fasting plasma glucose (FBG)=60-100 mg/dl and 2 hours post load glucose level <140 mg/dl. It is useful in identifying different types of glycosurias like renal glycosuria, alimentary glycosuria. It is helpful in detection of diabetes.

3.

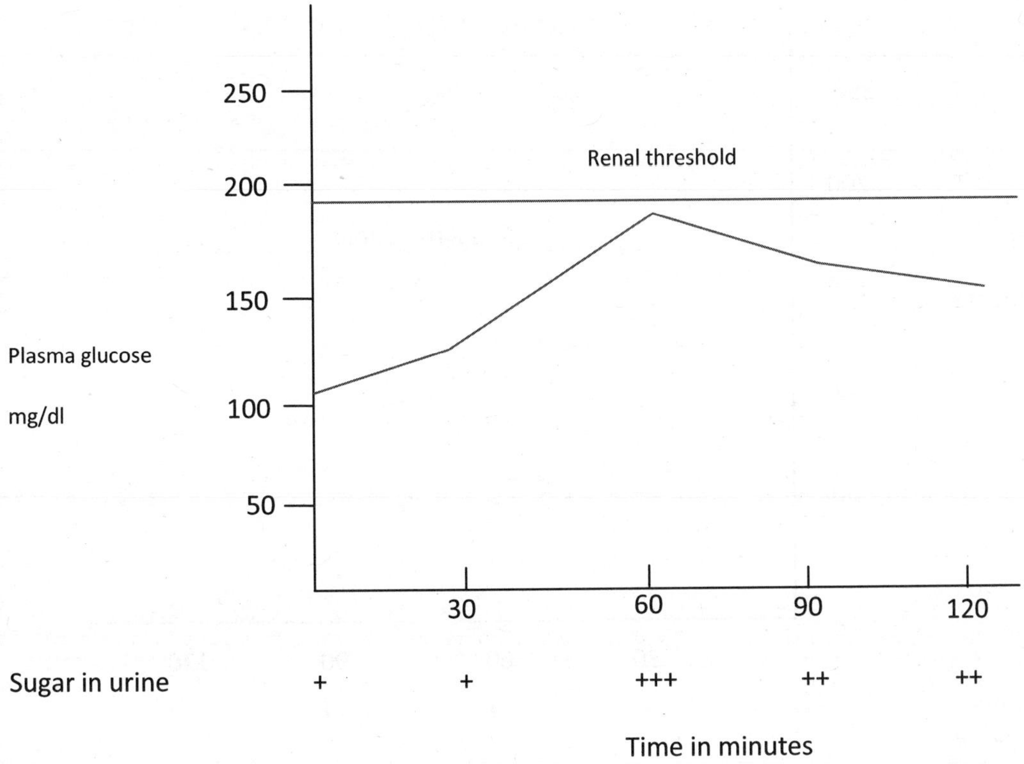

i. Identify the graph.

ii. Give the significance of this graph.

Answer

i. Oral glucose tolerance test curve indicating impaired glucose tolerance (IGT).

ii. Fasting plasma glucose (FBG) between 100-126 mg/dl and 2 hours post load glucose level between 140-199 mg/dl.

4.

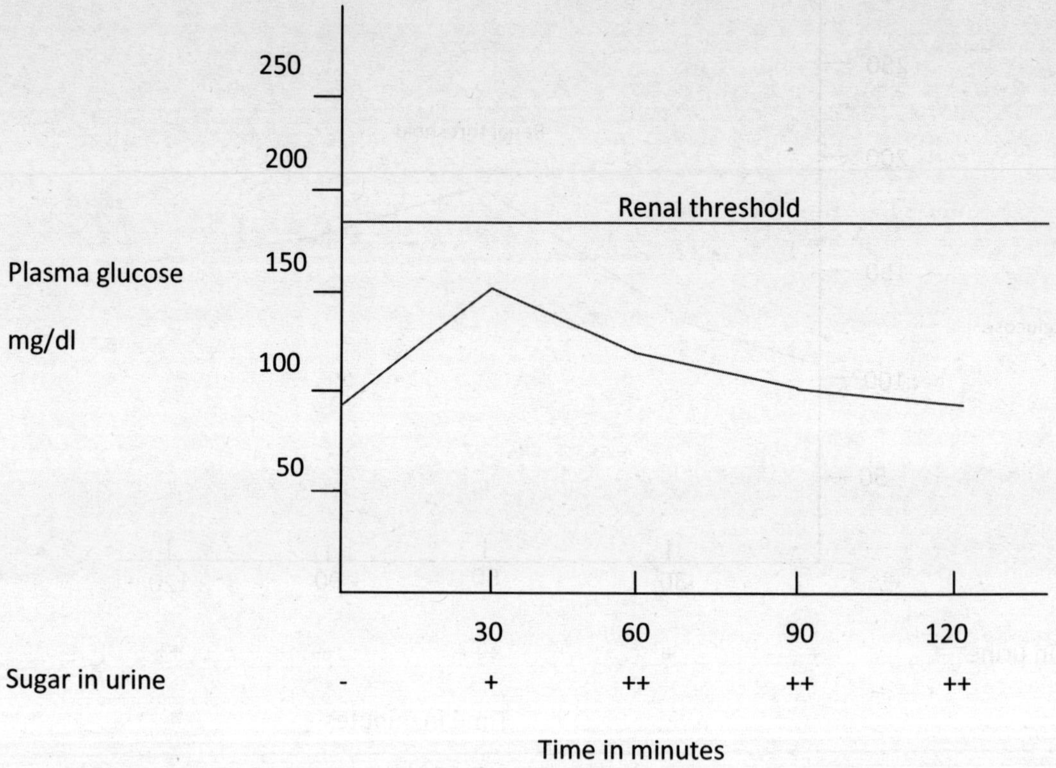

i. Identify the graph.

ii. Give the significance of this graph.

Answer

i. Oral glucose tolerance test curve indicating renal glycosuria.

ii. Though the blood glucose level is within the normal range, sugar appears in urine due to decrease of renal threshold for glucose.

5. i. Identify the disease.

ii. Mention the cause.

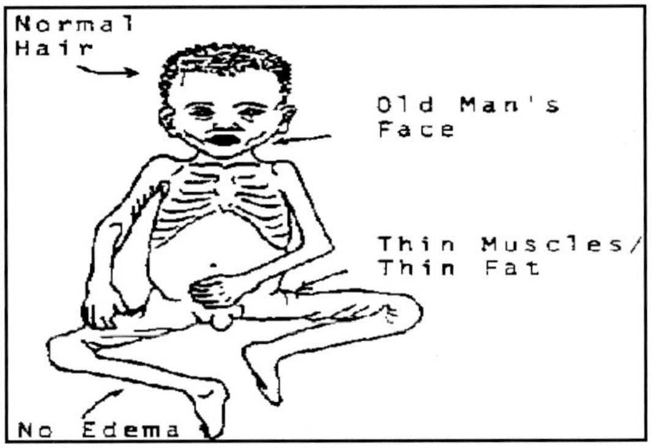

Answer

i. Marasmus

ii.Inadequate caloric intake

6. i. Identify the disease.

ii. Mention the cause.

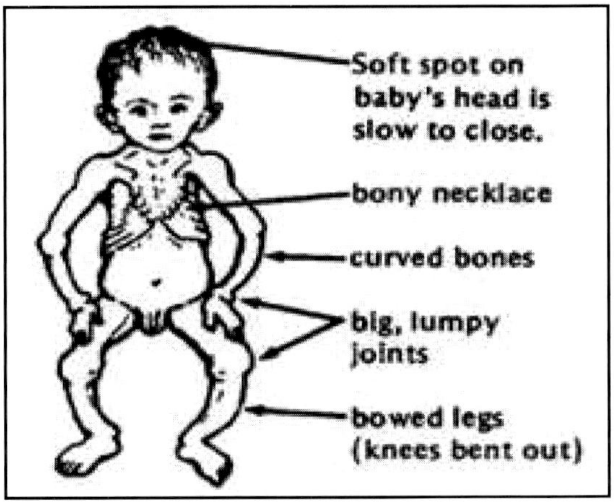

Answer

i. Rickets

ii. Vitamin D deficiency

7. i. Identify the disease

ii. Mention the cause

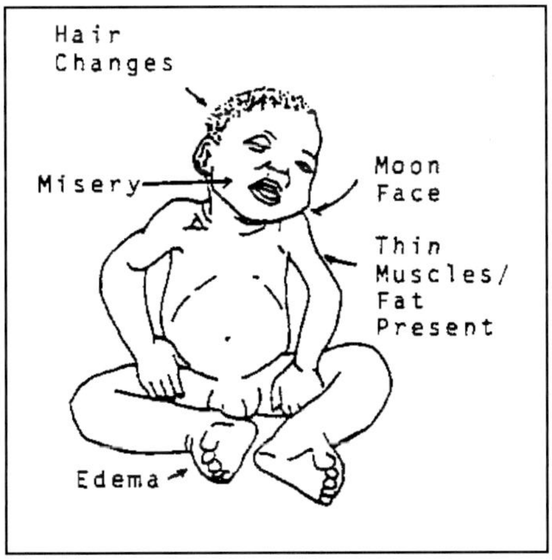

Answer

i. Kwashiorker

ii. Inadequate protein intake

8. i. Identify the disease

ii. Mention the cause

Answer

i. Pellagra

ii. Niacin deficiency

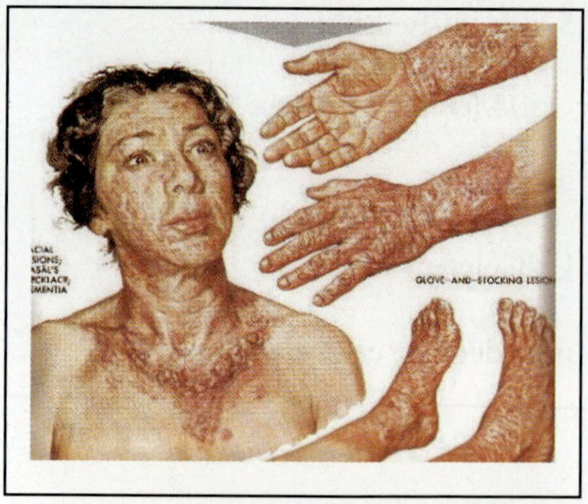

9. i. Identify the disease

ii. Mention the cause

Answer

i. Scurvy

ii. Vitamin C deficiency

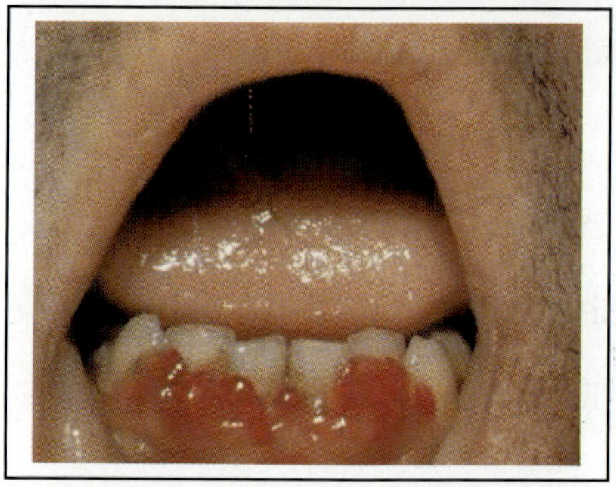

10. i. Identify the disease

ii. Mention the cause and treatment

Answer

i. Gout

ii. Excess catabolism of nucleotide. Treated by allopurinol

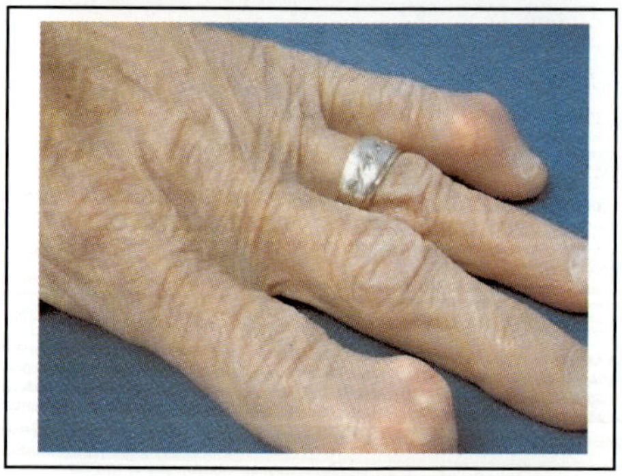

11. i. Identify the disease

ii. Mention the cause

Answer

i. Xerophthalmia

ii. Vitamin A deficiency

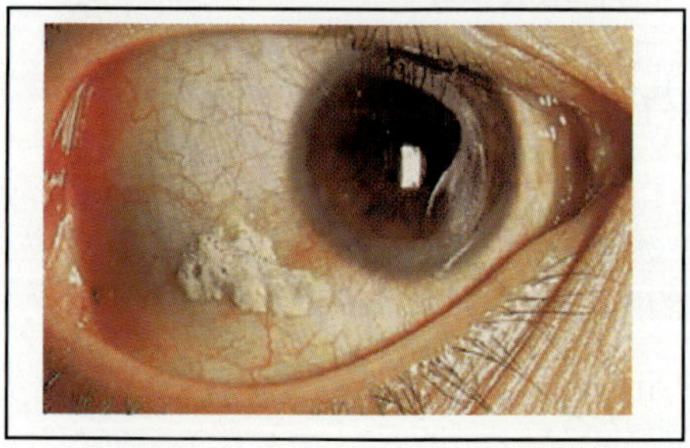

12.i. Identify the instrument.

ii. Mention the use.

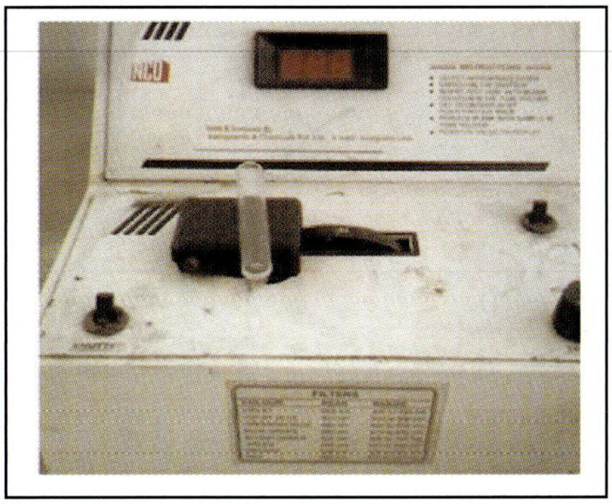

Answer

i. Colorimeter.

ii. Measurement OD in quantitative experiments.

13.i. Identify the instrument.

ii. Mention the use.

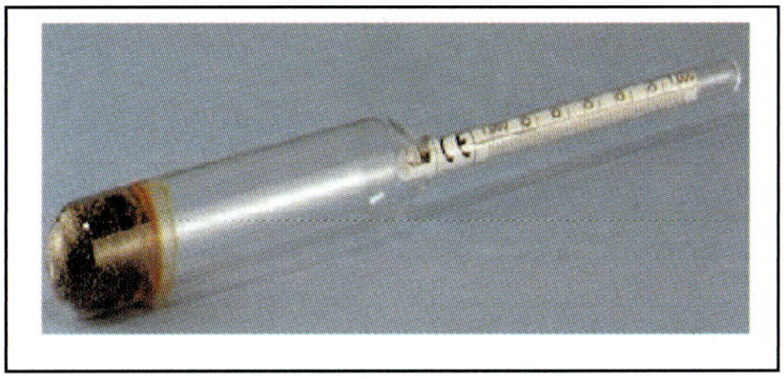

Answer

i.Urinometer.

ii. Measurement of specific gravity of urine.

14.i. Identify the instrument.

ii. Mention the use.

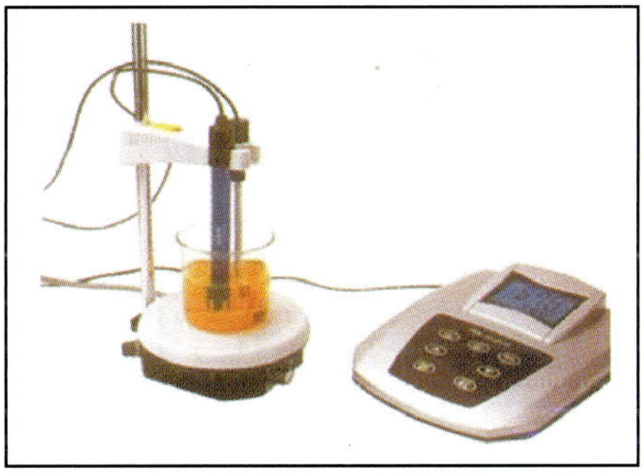

Answer

i. pH Meter.

ii. Measurement of pH of fluids.

15.i. Identify the instrument.

ii. Mention the use.

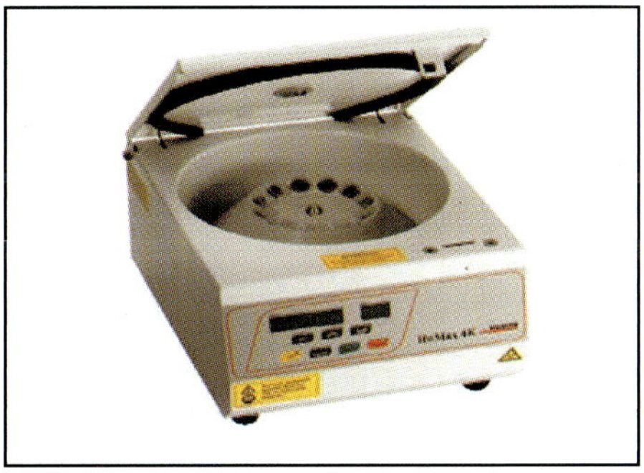

Answer

i. Centrifuge.

ii. Separation of solids from liquid suspension.

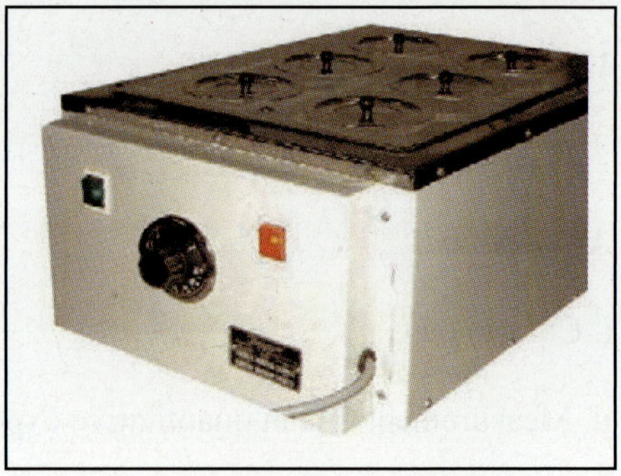

16.i. Identify the instrument.

ii. Mention the use.

Answer

i. Waterbath

ii. Provide heating environment for biochemical reactions

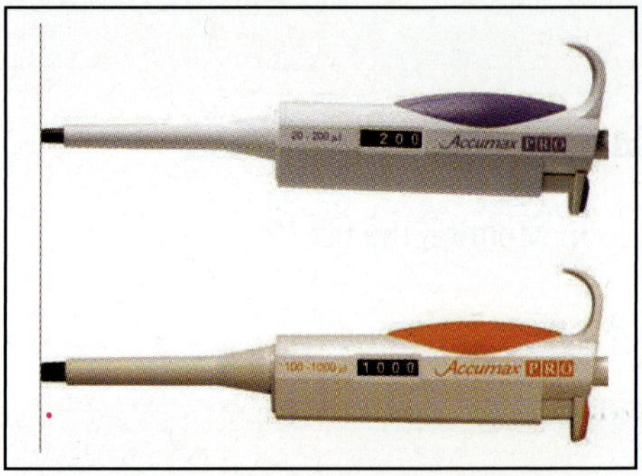

17.i. Identify the instrument.

ii. Mention the use.

Answer

i. Autopipette

ii. Transfer minute volume in accurate quantity

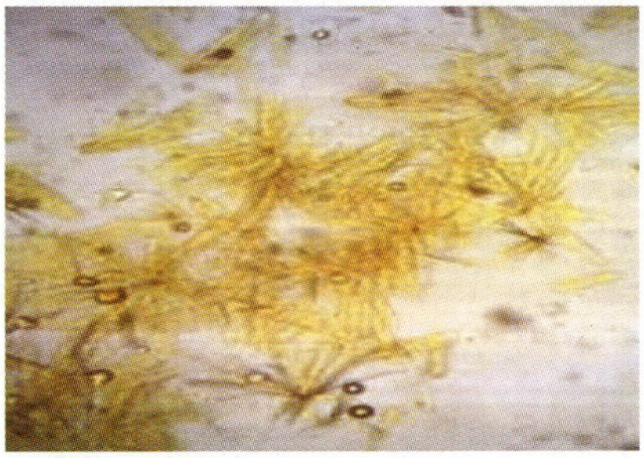

18.i. Identify the test.

ii. Name the compound giving this test.

Answer

i. Osazone test

ii. Maltose (sunflower shaped maltosazones)

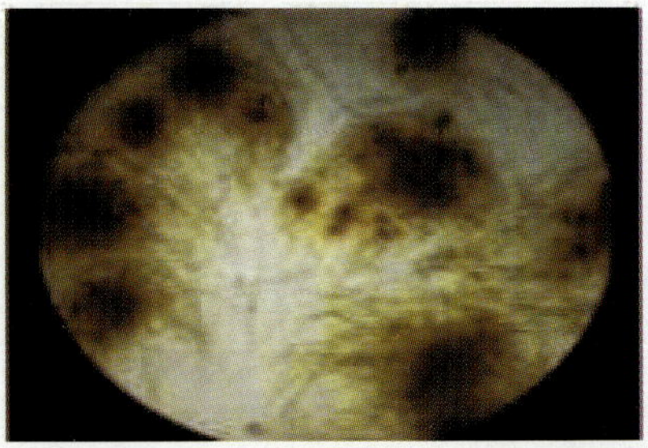

19.i. Identify the test.

ii. Name the compound giving this test.

Answer

i. Osazone test.

ii. Lactose (cotton ball shaped lactosazones).

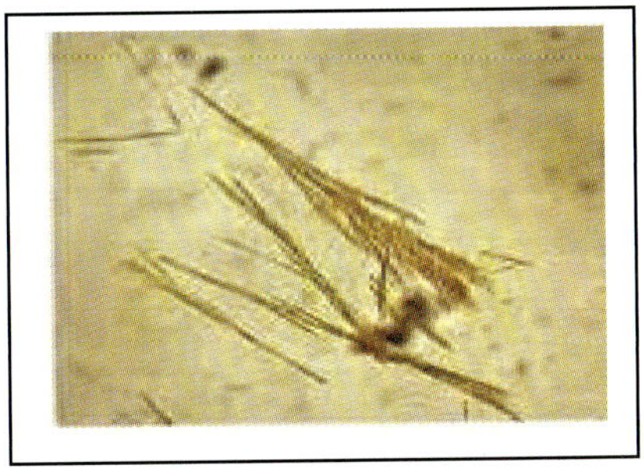

20.i. Identify the test.

ii. Name the compound giving this test.

Answer

i. Osazone test.

ii. Glucose, Fructose, Mannose (needle shaped osazones).

21.i. Identify the test.

ii. Give the use.

Answer

i. Barfoed's test.

ii. Differentiating monosaccharides from reducing disaccharides.

22.i. Identify the test.

ii. Name the compound giving this test.

Answer

i. Xanthoproteic test.

ii. Aromatic amino acids (tryptophan, tyrosine, phenylalanine).

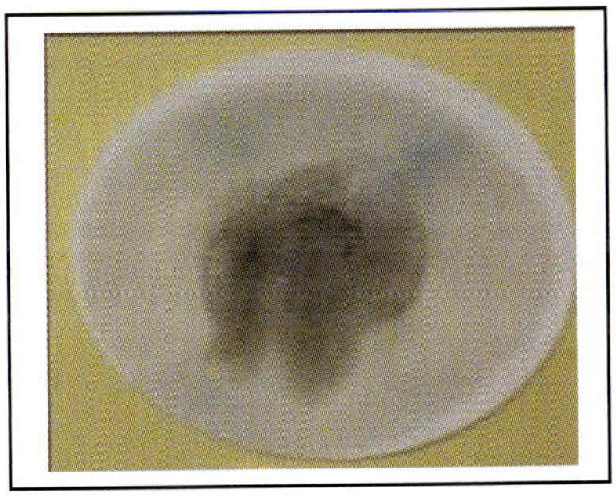

23.i. Identify the test.

ii. Mention the use.

Answer

i. Fouchet's test.

ii. Detection of Bile pigment in urine (Bilirubin).

24.i. Identify the structure.

ii. Give three compounds formed from it.

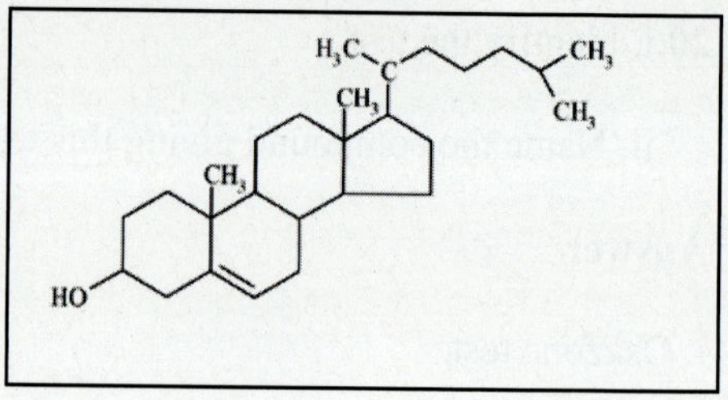

Answer

i. Cholesterol.

ii. Steroid hormones, Vitamin D, Bile salts.

25.i. Identify the structure.

ii. Mention the bond present.

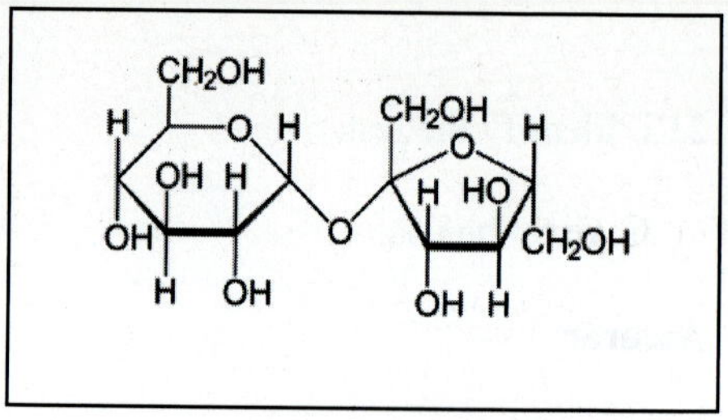

Answer

i. Sucrose (cane sugar).

ii. α1, β2 glycosidic bond.

26.i. Identify the structure and give its types.

ii. Give the level of organization.

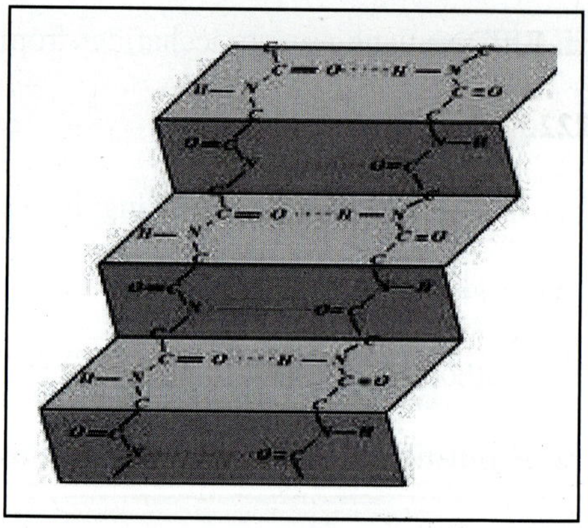

Answer

i. β-pleated sheets (Types: Parallel and anti-parallel)

ii. Secondary structure of protein

27.i. Identify the structure.

ii. Give the function.

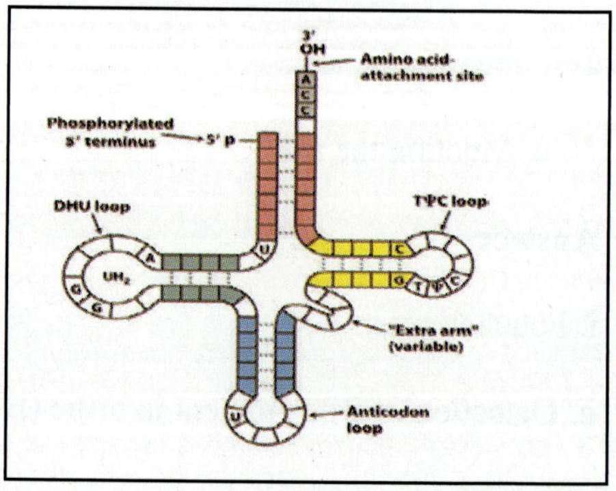

Answer

i. tRNA.

ii. Transfer of amino acid during protein synthesis.

28.i. Identify the graph.

ii. Give the example.

Answer

i. Lineweaver-Burk (LB) plot showing competitive inhibition.

ii. Treatment of methanol poisoning.

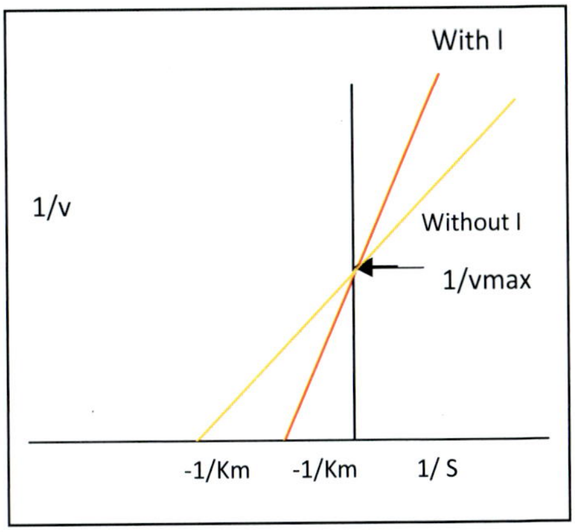

29.i. Identify the graph.

ii. Give the example.

Answer

i. Lineweaver-Burk(LB) plot showing non-competitive inhibition.

ii. Inhibition of enolase by fluoride for blood glucose estimation.

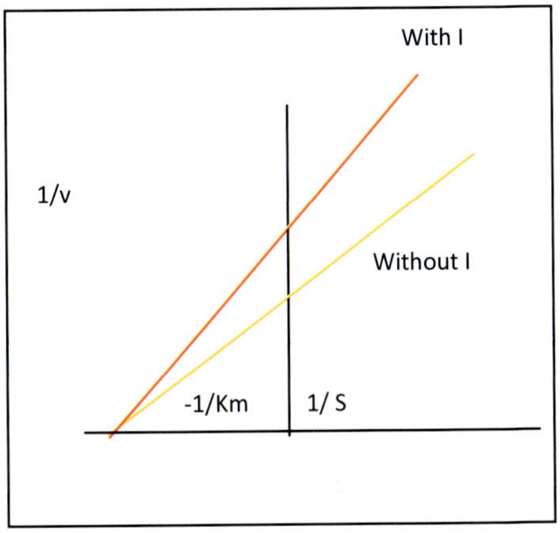

30.i. Identify the cycle.

ii. How many ATPs produced/cycle.

Answer

i. TCA (Tri carboxylic acid) cycle.

ii. 12 ATP (3 NADH=9 ATP, 1 $FADH_2$=2 ATP, 1 GTP).

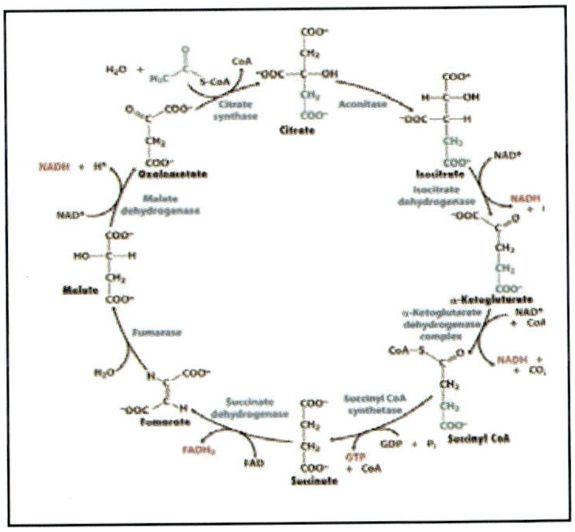

31.i. Identify the picture.

ii. Give the role.

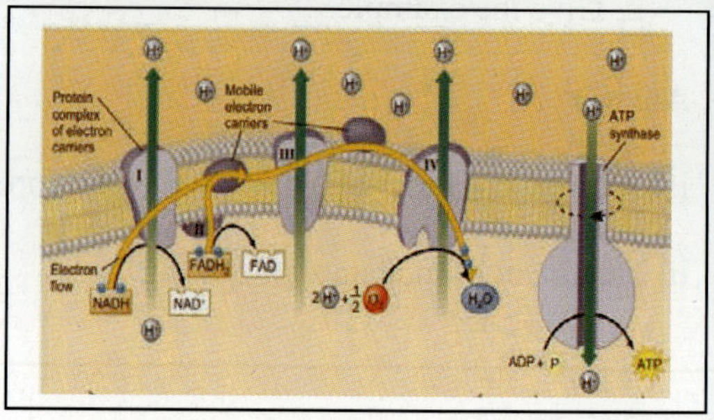

Answer

i. Electron transport chain.

ii. Oxidative phosphorylation (ATP production).

Appendix V

VIVA QUESTIONS

Carbohydrates

Molisch's Test

1. Why Molisch's test considered general test for carbohydrates?
   - It is given by all carbohydrates starting from pentose.
2. What is other general test for carbohydrates?
   - Anthrone test (gives blue color with carbohydrates).
3. Do Trioses and Tetroses give Molisch's test positive?
   - No, because furfural derivative formed in the reaction is 5-carbon compound.
4. How many molecules of water are lost during furfural formation?
   - Three (3).
5. Name the furfurals of pentose and hexose?
   - Pentose → furfural.
   - Hexoses → hydroxymethyl furfural.
6. Give the name of Molisch's reagent?
   - 1% α-naphthol prepared in ethanol.
7. Give examples of carbohydrates.
   - Polysaccharides —— Starch and glycogen.
   - Disaccharides —— Maltose and lactose.
   - Monosaccharides —— Glucose, fructose, mannose.

Iodine Test

1. How can you distinguish between starch, dextrin and glycogen on the basis of iodine test?
   - Depending on the color developed they can be identified.

     Starch → Blue color
     Glycogen → Reddish color
     Dextrin → Violet color
2. Why does starch gives blue color and glycogen gives reddish brown color with iodine?
   - Starch and glycogen contains two components namely amylose and amylopectin. Iodine is trapped in helical structure of amylose producing the color. In comparison to starch, glycogen is highly branched and has more helical structure, so it holds more iodine giving reddish brown color.

3. Why blue color of starch-iodine complex disappear on heating and reappear on cooling?
   - On heating, the helical structure of amylose of starch distorts causing release of trapped iodine, so the blue color changes to colorless. On cooling, there is reformation of helical structure and retrapping of iodine molecule to produce color again. Color developed again may not be as intense as before, since some amount of iodine may get vaporized on heating.

Benedict's Test

1. Does starch give Benedict's test positive?
   - No, it is because starch is a large molecule with only one reducing end, so it's reducing activity is negligible in comparison to monosaccharides and disaccharides.
2. What is the use of Benedict's test?
   - Differentiating reducing sugars from non-reducing sugar.
3. What are reducing sugars? Give examples.
   - Sugars having free aldehyde or keto group are known as reducing sugars, e.g., Glucose, Fructose, Lactose, Maltose, Mannose, Galactose etc.
4. Give the composition of Benedict's reagent.
   - Na_2CO_3 → provides alkaline medium.
   - $CuSO_4.5H_2O$ → provides cupric (Cu^{2+}) ions.
   - Sodium citrate → stabilizing agent. Protects Cu^{2+} ions from precipitation.
5. Why Benedict's test is known as semi-quantitative test?
   - When fixed volume of Benedict's reagent (5 ml) and sugar solution (8 drops) is added quantity of sugar present in the solution can be tentatively determined, depending upon the color developed.

Green upto → 0.5%

Green ppt → 0.5-10%

Yellow ppt → 1-1.5%

Orange → 1.5-2%

Brick red → More than 2%

6. What is Enediol?
   - Enediol is strong reducing agent formed by tautomerization of reducing sugars. Tautomerization means shifting of hydrogen atom from one carbon to other in same molecule.

Barfoed's Test

1. What is the use of Barfoed's test?
   - Differentiating monosaccharides from reducing disaccharides.
2. Give the composition of Barfoed's reagent.
   - Glacial acetic acid → provides acidic medium.
   - Cupric acetate→ provides cupric ions.
3. What is the difference between Benedict's and Barfoed's test?

	Benedict's	**Barfoed's**
Use	Given positive by all reducing sugars	Given positive by monosaccharides only
Medium	Alkaline	Acidic
Source of Cu^{2+}	$CuSO_4$	Cupric acetate

Seliwanoff's Test

1. Why we perform Seliwanoff's test?
   - To differentiate ketohexoses from aldohexose. It is given positive by ketoses. e.g., fructose.
2. What is the test for ketopentoses?
   - Bial's test. It includes orcinol, conc. HCl and ferric chloride. Pentoses are dehydrated by concentrated HCl to furfurals which then react with orcinol and iron to form bluish colored product.
3. Give the composition of Seliwanoff's reagent.
   - Resorcinol and conc.HCl.
4. Sucrose is non-reducing sugars but gives Seliwannof's test positive. Why?
   - It is because Seliwanoff's reagent contains concentrated HCl. The acidity is sufficient to hydrolyze sucrose to glucose and fructose. The resulting fructose will then give Seliwanoff's test positive.

Osazone Test

1. What is the shape of osazone crystals formed by glucose, fructose, mannose, galactose, lactose and maltose ?
   - Glucose, fructose, mannose →Needle shaped
   - Lactose → Cotton ball shaped
   - Maltose → Sunflower shaped
   - Galactose→ Rhombic plate shaped
2. Why glucose, fructose and mannose give same osazone crystals (needle shaped)?
   - Osazone reaction takes place in first and second carbon atom of reducing sugar. Any sugar differing at these two carbon atoms produce same osazone crystals as the difference is

masked by the reaction. Glucose, fructose and mannose differ at C-2. Rest structure after second C is same for these 3 sugars, so they produce same osazone crystals.

SUCROSE

1. Why sucrose is non-reducing sugar?
   - Sucrose is made up of fructose and glucose linked with α1, β2 glycosidic linkage. Both glucose and fructose are reducing sugars but during the bond formation the free end of these sugars are utilized, so sucrose is non-reducing sugar as it does not contain free aldehyde or keto group.
2. Why sucrose is known as invert sugar?
   - Sugar that contains mixture of equal part of glucose and fructose is known as invert sugar. Sucrose is dextrorotatory (rotates plane of polarization towards right/clockwise: +66.5°). On hydrolysis, sucrose gives glucose and fructose. Glucose is dextrorotatory and fructose is levorotatory (rotates plane of polarized light towards left/ anticlockwise). The degree of levorotation caused by fructose is higher than dextrorotation caused by glucose. So sucrose that was dextrorotatory before hydrolysis is converted to levorotatory on hydrolysis. There is change in the rotation of optical path, so sucrose in known as invert sugar.
3. Why lactose is reducing sugar?
   - Lactose is made up of galactose and glucose (β-1,4). During bond formation free aldehyde end of galactose is utilized but that of glucose is free. It is due to free reducing end in glucose molecule, lactose is reducing in nature.

PROTEIN

Precipitation Reactions

1. What is the significance of protein precipitation?
   - Protein precipitation is used for separation of proteins from biological fluids (serum, CSF) before estimation of various biochemical parameters like glucose, creatinine, urea, uric acid etc. Precipitation reaction can also be used for separation of protein of interest from the mixture of protein, e.g., separation of globulin from plasma by half saturation with ammonium sulphate.
2. What is salting out?
   - It is precipitation of proteins from the solution by using solution of neutral salt like $(NH_4)_2SO_4$, Na_2SO_4, e.g., half saturation test.
3. What is salting in?
   - When solution with lower concentration of salt is added to the protein solution, it increases the solubility of the protein in the solution. This is known as salting in.

4. Why albumin requires full saturation with $(NH_4)_2SO_4$, whereas globulin is precipitated on half saturation with $(NH_4)_2SO_4$?
   - ➢ Amount of salt required for protein precipitation depends on the molecular weight, size and surface area of the protein molecule. Molecular weight is directly proportional to the size of protein molecule whereas surface area is inversely proportional to the molecular weight. More is the surface area; more amount of salt is required for protein precipitation.

     In comparison to globulin, albumin is smaller in size with lower molecular weight, but the surface area of albumin molecule is high so it requires higher salt concentration to be precipitated. It is precipitated by full saturation with neutral salts. Conversely, globulin is a large size protein with high molecular weight. It has less surface area, so it requires lower salt concentration. It is precipitated by half saturation with $(NH_4)_2SO_4$.

5. Why peptone is not precipitated even by full saturation with $(NH_4)_2SO_4$?
   - ➢ Peptone is a lower molecular weight protein. Its size is smaller than albumin, so it is not precipitated even by full saturation with $(NH_4)_2SO_4$.
6. What is the clinical significance of egg albumin?
   - ➢ It can be used as an antidote in acute lead poisoning or mercury poisoning. Egg albumin forms insoluble precipitate with lead and mercuric salt. This retards the absorption of this metal protein complex in stomach and intestine.
7. Why acetic acid is added in heat coagulation test?
   - ➢ To differentiate whether the white coagulum formed is due to protein or due to the presence of phosphates. If the coagulum thickens on addition of acetic acid, then it is due to protein. If the coagulum disappears on addition of acetic acid it is due to phosphates.
8. Why upper one third part is heated in case of heat coagulation test?
   - ➢ Upper 1/3rd part is heated because:
     - Lower 1/3rd part acts as a control.
     - When solution is heated, hot air moves from lower end to upper end, so if the solution is heated from the bottom, whole solution will turn white (due to coagulation).Therefore to take lower part as control and to aid in comparison, upper 1/3rd is heated.
9. What is isoelectric pH (pI)?
   - ➢ pH at which the protein contains equal amount of positive and negative charges. The molecule is electrically neutral e.g., pI of caesin is 4.6.
10. What is denaturation of protein?
    - ➢ It is unfolding of 3-D structure of protein. There is loss of 2°, 3° and quaternary structure, but 1° structure remains intact. Biological activity of protein is lost as 2° structures are required for the protein to be active.

Color Reactions of Proteins

1. Why Biuret test is called so?
   - Urea when heated at 180°C forms the compound biuret which gives the biuret test positive; hence the test got its name Biuret.
2. What is the requirement for biuret test to be positive?
   - At least two peptides bonds should be present for the biuret test to be positive, e.g., tripeptide (2 peptide bond, 3 amino acids).
3. What is the name of complex formed in Biuret test?
   - Copper coordination complex.
4. The solution gives Biuret test negative but Ninhydrin test positive. What does it mean?
   - The solution contains free α-amino acids but no peptide linkages.
5. What is the name of complex formed in ninhydrin test?
   - Ruhemann's purple.
6. Why does proline and hydroxyproline give yellow color not purple colored complex with ninhydrin test?
   - Proline is an imino acid which does not contain α-amino group. Thus it does not give purple color with ninhydrin. Rather ninhydrin forms yellowish brown colored condensation product with proline.
7. What is the use of xanthoproteic test?
   - It is to identify aromatic amino acids (tyrosine, tryptophan, phenylalanine).
8. Among aromatic amino acid which one is essential and non-essential?
   - Tryptophan and phenylalanine ⟶ Essential.
   - Tyrosine ⟶ Non-essential.
9. When does tyrosine become essential?
   - Tyrosine is non-essential amino acid but becomes essential in case of phenylketonuria. In phenylketonuria, tyrosine cannot by synthesized from phenylalanine due to deficiency of phenylalanine hydroxylase. Thus tyrosine becomes essential (should be supplied externally).
10. Why tryptophan gives aldehyde test positive?
    - Hope-Cole's-Adam's aldehyde test is given positive by the compound containing indole group. Tryptophan contains indole group.
11. Name the group present in tyrosine, arginine and histidine.
    - Tyrosine ⟶ Hydroxyphenyl group (Million's and Pauly's test positive)
    - Arginine ⟶ Guanidino group (Sakaguchi test positive)
    - Histidine ⟶ Imidazole group (Pauly's test positive)

12. Test

Solution	Million's Test	Pauly's Test
A	Positive	Positive
B	Negative	Positive

What does it indicate?

- A ⟶ Tyrosine
- B ⟶ Histidine

13. Why methionine gives sulphur test negative though it is a sulphur containing amino acids?
    - Methionine contains thio-ether linkage which is not cleaved even on heating with strong alkali, so the organic sulphur present in methionine is not released. Hence sulphur test is negative for methionine.
14. Name semi-essential amino acids.
    - Arginine and Histidine.

LIPIDS

1. Triacylglycerol is called neutral lipid. Why?
   It does not have any charged or polar groups.
2. Name some polar and amphipathic lipids.
   Fatty acids, phospholipids, cholesterol are some polar lipids. Phospholipids contain both polar (phosphoric acid and nitrogenous compounds) and non-polar (acyl chain) groups. So they are called as amphipathic lipids.
3. How can you make comparison between animal fat and plant fat?

   Animal fat
   - Usually contain saturated fatty acid.
   - Contain cholesterol so are atherogenic in nature.
   - Are solid at room temperature.
   - Sources: Butter, egg, milk etc.

   Plant fat
   - Contain unsaturated fatty acid.
   - Does not contain cholesterol so non-atherogenic in nature.
   - Are liquid at room temperature.
   - Sources: Plant oils.

4. Compare fat and oil.
   Fats are solid at room temperature and oils are liquid at room temperature.
5. Give example of biological detergent.
   Bile salts, E.g., sodium glycocholate and sodium taurocholate. They help in emulsification and digestion of fat.
6. Soaps are used as cleansing agents. Why?
   They remove lipid particles by emulsification and thus remove dirt particles and stains present in them.

Urine Analysis

1. What are the causes of abnormal color of urine?
   - Dark Yellow ⟶ Jaundice
   - Black ⟶ Alkaptonuria
   - Red ⟶ Hemoglobinuria/Hematuria etc.
2. What are the causes of the abnormal odor of urine?
   - ➢ Mousy ⟶ Phenyl ketonuria
   - ➢ Mapple syrup/burnt sugar ⟶ Mapple syrup urine disease
   - ➢ Rancid ⟶ Tyrosinemia
   - ➢ Fruity ⟶ Acetone
3. What is chyluria?
   - ➢ Presence of fat droplets in urine giving the milky appearance.
4. What is oliguria and anuria? Give causes.
   - ➢ Excretion of <600 ml of urine per day is oliguria. Causes may be shock, excessive diarrhea, vomiting and sweating.
   - ➢ Anuria means excretion of <100 ml of urine per day. Causes may be renal failure or shock.
5. What do you mean by glycosuria? Mention types and causes.
   - ➢ Presence of reducing sugar in urine is called glycosuria. Types may be:
     a. Physiological glycosuria: Lactation, Pregnancy.
     b. Pathological glycosuria: Diabetes
     c. Renal glycosuria: Due to decreased renal threshold for glucose.
     d. Alimentary glycosuria: Excessive carbohydrate intake.
6. What is renal threshold for glucose? What does it mean?
   - ➢ Renal threshold for glucose is 180 mg/dl. It means unless the blood glucose level is more than 180 mg/dl, it does not appear in urine.
7. How can one differentiate between physiological and pathological glycosuria?
   - ➢ It can be done by blood glucose estimation. In physiological glycosuria blood glucose is normal but sugar appears in urine due to increased GFRm, e.g., pregnancy and lactation.

In pathological glycosuria, blood glucose is high (more than renal threshold), so it appears in urine, e.g., Diabetes mellitus.

8. The level of blood glucose is normal but there is presence of glucose in urine. What does it indicate?
   - Renal glycosuria or alimentary glycosuria.
9. What is proteinuria? What are the causes?
   - Excretion of protein in urine is proteinuria. Causes may be nephrotic syndrome, renal failure, diabetic nephropathy etc.
10. Which protein is most commonly excreted?
    - Albumin.
    - Normal excretion : <30-150 mg/day.
    - Microalbuminuria (excretion of less amount of albumin):150-300 mg/day.
    - Proteinuria: >300 mg albumin/day.
11. What is multiple myeloma? Which protein is excreted in urine?
    - It is the cancer of plasma cell.
    - Bence Jones protein (γ-globulin light chain) is excreted in urine.
12. What is distinguishing feature of Bence Jone's protein?
    - When Bence Jones protein is heated at 50-60°C, it coagulates and turbidity appears but as heating is continued up to 100°C, the turbidity disappears. It reappears again on cooling. Thus one can suspect multiple myeloma by doing simple heat coagulation test too.
13. What is the electrophoretic pattern of serum or urine proteins in multiple myeloma?

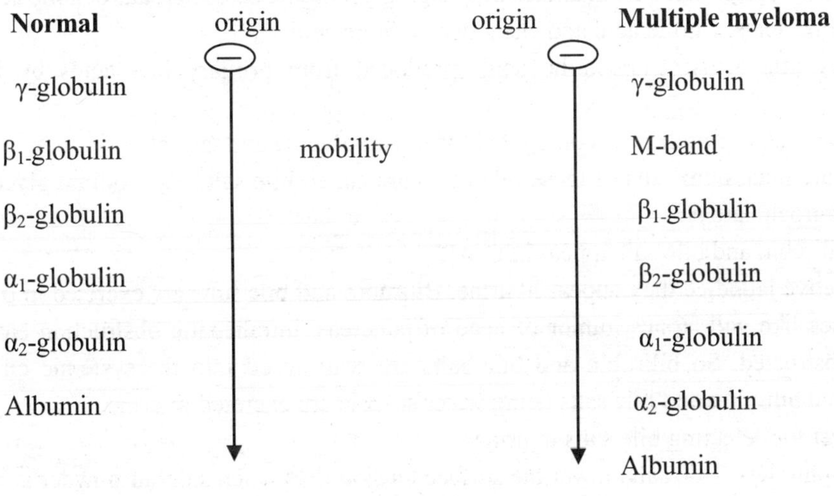

Extra M-band is seen in multiple myeloma (between γ-globulin band and β-globulin band). M-band contains L-chains of immunoglobins .

14. Give examples of ketone bodies. What is the clinical significance?
    - ➢ Acetone, acetoacetate, β-hydroxybutyrate.
    - ➢ They are synthesized in liver from acetyl CoA. Level of ketone bodies is increased in prolonged starvation and uncontrolled diabetes.
    - ➢ Ketone bodies lead to ketoacedosis.
15. Which ketone body is most potent in causing acidosis?
    - ➢ β-hydroxybutyrate.
16. What is the test to detect ketone bodies in urine?
    - ➢ **Rothera's test:** It is given positive only by acetone and acetoacetate but not by β-hydroxybutyrate because β-hydroxybutyrate contains hydroxyl group in place of keto group.
17. What is normal level of blood ketone bodies and their renal threshold?
    - ➢ Normal blood ketone bodies: < 1 mg/dl.
    - ➢ Renal threshold for ketone bodies: 3 mg/dl.
18. Bilirubin is yellow colored but there is development of green color in Fouchet's test, why?
    - ➢ Bilirubin present in urine is adsorbed on $BaSO_4$ and is separated on filter paper. On addition of Fouchet's reagent, it is oxidized by $FeCl_3$ of the reagent to biliverdin which is green in color.
19. Why $BaCl_2$ and $MgSO_4$ are added in Fouchet's test?
    - ➢ $BaCl_2$ and $MgSO_4$ react to form white precipitate of $BaSO_4$. Bilirubin in urine is then adsorbed on the surface of $BaSO_4$ that can be precipitated on filter paper.
20. What are bile salts and bile acids?
    - ➢ Bile acids are synthesized from cholesterol, e.g., glycocholic acids and taurocholic acids.
    - ➢ Primary bile acids: Cholic acid and chenodeoxycholic acid.
    - ➢ Secondary bile acids: Lithocholic acid (produced from primary bile acids by intestinal bacteria).
    - ➢ These bile acids are present in conjugated form with glycine and taurine.
    - ➢ Sodium and potassium salts of these bile acids are called bile salt, e.g., sodium glycocholate, sodium taurocholate etc.
21. When does bilirubin and bile salt appear in urine?
    - ➢ In obstructive jaundice they appear in urine. Bilirubin and bile salts are excreted in bile but in some cases like gall stones, tumor of head of pancreas, intrahepatic obstruction etc flow of bile is obstructed. So bilirubin and bile salts are reabsorbed into the systemic circulation. Conjugated bilirubin and bile salts being water soluble are excreted in urine.
22. What is the test for detecting bile salts in urine?
    - ➢ Hay's sulphur test: Bile salts lower the surface tension thus when sulphur powder is sprinkled over the urine sample, it settles at the bottom of the tube.
23. What is the importance of Benzidine test?
    - ➢ Detect blood in urine. It uses peroxidase property of heme.

24. Give conditions in which there is presence of blood in urine.
Glomerulonephritis, renal stones, UTI, malignancies, tuberculosis, trauma of kidney etc.

Colorimetry

1. State Beer's law.
   - ➢ Absorbance α concentration (if pathlength travelled by light is kept constant):
2. State Lambert's law.
   - ➢ Absorbance is α pathlength (if concentration of solution is kept constant).
3. Why complementary filters are used in colorimeter?
   - ➢ When white light passes through a colored solution, a portion of the light is absorbed by the coloring substance (also called chromogen) and the rest is transmitted. The extent to which different components of white light are absorbed by a solution depends upon the nature of the chromogen. E.g., if the chromogen forms red colored solution, the absorption of red component is minimum whereas the other components are absorbed to the larger extent. The complementary color, blue in this case is absorbed to the greatest extent. In colorimetry, it is preferable to use monochromatic light of a specific wavelength which is absorbed maximally by the chromogen being measured. So complimentary filters are used to get absorption maxima.
4. Mention part of colorimeter.
   - Light source → Filament lamp (tungsten lamp)
   - Filter → Selection of wavelength
   - Cuvette → Sample container tube
   - Photocell → Gives reading
5. Give differences between colorimeter and spectrophotometer.

	Colorimeter	Spectrophotometer
Light solution	Filament lamp (tungusten lamp), only colored solution used	Deuterium lamp works on both colored and colorless solution
Working range	Visible	UV, Visible, IR
Wavelength selection	By filter	By diffraction grating or prism
Cuvette	Made up of borosilicate glass	Made up of quartz glass
Sensitivity	Low	High
Cost	Cheap	Expensive

Blood Sugar

1. Give normal range of blood glucose.
   - ➢ Fasting: 60-100 mg/dl.
   - ➢ Post prandial:100-140 mg/dl.
   - ➢ Random: 60-140 mg/dl.

2. What do you mean by fasting, PP and random blood sugar?
   - Fasting: Sample collected after overnight or 12 hours of fast.
   - Post prandial: Sample collected 2 hours after carbohydrates meal.
   - Random: Sample taken at any time irrespective of food intake.
3. What are the blood samples used for blood sugar estimation?
   - Venous sample (most common):
   - Arterial sample
   - Capillary sample (used for self testing using glucometer). Capillary sample glucose values are 2-5 mg/dl more than venous sample values.
4. What are the anticoagulants used for blood sample collection for glucose estimation?
   - Potassium oxalate and sodium fluoride in the ratio 3:1.
   - Sodium fluoride is used to inhibit glycolysis. RBCs can utilize glucose even after sample collection resulting in decrease in glucose level by 5-10 mg/dl per hour, but addition of fluoride prevents this by inhibiting enolase enzyme of glycolysis.
5. What is the method for blood glucose estimation, which one is the method of choice?
   - O-Toluidine method
   - Folin Wi method
   - Nelson Somogyi method
   - King Astoor method

   } Chemical method

   - GODPOD method (glucose oxidase peroxidase)

   GODPOD is method of choice as it is specific and enzymatic method.
6. Why other methods of glucose estimation are non-specific?
   - ➢ Blood glucose estimation by chemical method is based on the reducing property of glucose, but the reaction can also be interfered by other reducing substances present in blood, e.g., ascorbic acid, uric acid, glutathione etc. In addition to glucose, these substances can also react with the reagent to produce color, thereby giving higher results, so the chemical methods are non-specific.
7. What is the method of choice among chemical methods?
   - ➢ **O-Toluidine method**

   O-Toluidine reacts with aldohexoses in presence of glacial acetic acid to produce green color. Glucose is major aldohexose in blood, so O-toluidine reacts predominantly with glucose but not with other reducing substance in blood. Hence it gives glucose values close to the true blood glucose value (fairly specific), but main drawback is that O-toluidine is not specific like GODPOD method.
8. What is true glucose?
   - ➢ It is the value of blood glucose estimated by enzymatic method.
9. Give the diagnostic criteria for diabetes.

- Fasting glucose ≥ 126 mg/dl
- PP glucose ≥ 200 mg/dl
- Random glucose ≥ 200 mg/dl with symptoms like polyuria, polydypsia and polyphagia.

  If any one of these three criteria (a,b,c) is satisfied, then the person is considered diabetic.

10. What are causes of hyper and hypoglycemia?
    - ➢ Hypoglycemia: Diabetes mellitus, hyperactivity of pituitary and adrenal gland, encephalitis, prolonged use of steroids.
    - ➢ Hypoglycemia: Insulin overdose

      Hyperthyroidism

      Starvation, glycogen storage diseases.
11. How is level of blood sugar regulated?
    - ➢ It is regulated mainly by two hormones insulin and glucagon. Insulin is hypoglycemic hormone (decrease blood glucose level).When blood glucose level increases insulin is secreted. It lowers blood glucose by activating pathways that utilizes glucose, e.g., glycolysis, glycogenesis. In contrast to insulin, glycogen is hyperglycemic hormone (increase blood glucose).When blood glucose level is low, glycogen is secreted. It increases blood glucose by activating glycogenolysis, gluconeogensis and inhibiting glycolysis.
12. How diabetes mellitus is monitored?
    - ➢ By regular measurement of HbA_1C (glycosylated hemoglobin).

      Normal: 4-6%

      Good control: <6%

      Average: 6-8%

      Bad control: > 9%
13. What is OGTT? Give the significance?
    - ➢ **Oral glucose tolerance test**

      It is used to diagnose diabetes, IGT and different types of glycosuria (renal and alimentary).

Creatinine

1. What is creatinine? Give the normal level of serum creatinine?
   - ➢ It is end product of muscle protein metabolism.
   - ➢ Creatinine is formed from creatine by dehydration and spontaneous cyclization.

Glycine + Arginine + Methionine ⟶ Creatine

Creatine ⟶ Creatine— Phosphate

↓ H_2O

Creatinine + Pi

Normal range: 0.6-1.5 mg/dl (male) and 0.3-1.1 mg/dl (female)

2. What is the importance of estimating serum creatinine?
   ➢ It is indicator of renal function. Serum level increases in renal disorders.
3. What is the method for creatinine estimation?
   ➢ Jaffe's method: Orange colored complex is formed known as creatinine picrate (Janovsky complex).
4. Which one is better marker of kidney function (urea/creatinine)? Why?
   ➢ Creatinine: It is because serum creatinine is not influenced by factor like dietary protein intake, tubular reabsorption and secretion.
5. What is clearance? Give normal value.
   ➢ Clearance: Volume of plasma which is cleared of that amount of substance which is excreted by kidneys in 1 minute.
   Clearance = U × V/ P where U=concentration of substance in urine (mg/dl), P=concentration of substance in plasma (mg/dl), V= volume of urine in ml/minute.
   Normal range:
   - 95-140 ml/min (male)
   - 85-130 ml/min (female)
   - Average:125 ml/min
6. What are the clearance tests? Which one is the ideal one for GFR estimation?
   - Urea clearance
   - Creatinine clearance
   - Inuline clearance
   - Inuline is ideal substance for GFR estimation because of following reasons:
     a. It is neither produced nor metabolized in body.
     b. Freely filtered by glomerulus.
     c. It is not reabsorbed or secreted by kidney.
7. Though ideal for estimating GFR, why inulin clearance is not indicated to assess renal function?
   ➢ It is not endogenous compound and should be administered externally.
8. Which one is better creatinine clearance or urea clearance?
   ➢ Creatinine clearance is better marker for renal function because:
   - Creatinine is freely filtered by glomerulus.
   - Reabsorbed and secreted in negligible amount by renal tubules.
   - It is not affected by dietary protein intake.
   - Clearance value is almost close to GFR.
   - But urea clearance if affected by dietary protein intake. 40% of filtered urea is reabsorbed by renal tubule, so urea clearance is less than GFR.
9. What is normal excretion of creatinine?
   ➢ 1.5-2.0 gm/day (male)
   ➢ 0.8-1.5 gm/day (female)

10. What are the limitations of Jaffe's method? How can we overcome it?
   - Jaffe's method lacks specificity (not specific method). Only 80% of color produced is due to creatinine and rest 20% is due to reaction of other substance in serum (glucose, urea, uric acid etc.) with alkaline picrate.

     This problem can be overcome by removing these interfering substances using Lloyd's reagent (hydrated ammonium silicate). Llyod's reagent adsorbs creatinine in acidic medium and precipitate is collected by centrifugation. The pellet (precipitate) obtained by centrifugation is resuspended in alkaline medium to obtain creatinine which can be then estimated.

Urea

1. What is urea? Give normal range.
   - Urea is end product of protein and amino acids metabolism.

     Normal range: 20-40 mg/dl.
2. What are the methods to estimate blood urea?
   - Diacetylmonoxime method (chemical)
   - Urease method (enzymatic)
3. What is uremia?
   - Increase in blood urea level.
4. What is azotemia?
   - Increases in level of nitrogen content of blood.
5. What are the causes of uremia?
   - Pre renal causes (vomiting, dehydration, shock).
   - Renal causes (renal failure and glomerulonephritis).
   - Post renal causes (urinary tract obstruction, prostate enlargement etc.).
6. What is BUN?
   - Blood urea nitrogen:

     BUN= urea/2.14. The factor 2.14 is obtained as molecular weight urea/weight of nitrogen, i.e., 60/28=2.14.

     Normal range: 9-21 mg/dl in adults and 5-18 mg/dl in children.
7. What are standard and maximal urea clearance?
   - Based on flow of urine, renal function may be graded as:
     - Maximal urea clearance.
     - Standard urea clearance.
   - If flow of urine is 2 ml/min or more, it is called maximal urea clearance (UV/P).
   - If flow of urine is less than 2 ml/min, it is called standard urea clearance ($U\sqrt{V}/P$).
8. What is the significance of doing clearance test over the measurement of serum urea or creatinine?

- Clearance test indicate the earlier stage of renal damage at which blood urea or creatinine are likely to be normal.

Uric Acid

1. What is uric acid?
   - Uric acid is end product of purine nucleotide metabolism.
     Normal range: 3-7 mg/dl
2. Why uric acid is higher in males?
   - Due to higher muscle mass.
3. What are the methods to estimate serum uric acid?
   - Caraway's method (chemical method)
   - Uricase method (enzymatic method)
4. What is the influence of pH on uric acid?
   - Uric acid behaves as dibasic acid. It has two pK values: 5.75 & 9.8.
     At pH 5.75 and less, uric acid is present in unionized form (acidic form). At pH above 5.75, uric acid is present in ionized form (salt form i.e. monosodium urate).
     In urine due to acidification in renal tubules, pH decreases below 5.75. Thus in urine uric acid is excreted as such (i.e in the form of uric acid). In blood pH is 7.45, slightly alkaline (more than 5.75) so uric acid is present in the form of monosodium urate.
     Solubility of uric acid is 10 times less than that of monosodium urate.
5. What is hyperuricemia and gout?
   - Hyperuricemia: Increase in level of uric acid in blood.
   - Gout: Gout is the disorder resulting from hyperuricemia. It is characterized by arthritis and renal stones. Arthritis is due to deposition of sodium urate crystals in synovial fluid. These monosodium urate crystals deposited is known as tophi.
     Renal stones are due to deposition of uric acid.
6. What is the chemical nature of renal stones and tophi?
   - Renal stones are of uric acid. It is because glomerular filtrate of plasma in acidified in renal tubule causing decrease of pH. This leads to conversion of monosodium urate to uric acid. Uric acid is less soluble so it tends to deposit and results in renal stones.
   - Tophi are the monosodium urate crystals deposited in synovial fluid. Synovial fluid is ultrafiltrate of plasma whereas pH is around 7.45 (slightly alkaline). At this pH uric acid remains in the form of monosodium urate.
7. What are the causes of hyperuricemia?
   - Renal failure.
   - Gout.
   - Leukemia.
   - Polycythemia etc.

8. What are the types of gout?
   I. **Primary gout:** The causes are:
      - HGPRT deficiency.
      - PRPP synthatase over activity.
      - Adenosine deaminase deficiency.
      - G-6 phosphatase deficiency.
   II. **Secondary gout:** The causes are:
      - Leukemia.
      - Psoriasis.
9. Urine of gouty patient should be made alkaline. Why?
   - ➢ To enhance increased urinary excretion and to minimize the precipitation and deposition of uric acid in the form of renal stones.
   - ➢ At alkaline pH, uric acid remains in the form of monosodium urate that is 10 times more soluble than uric acid. It facilitates the excretion.
10. What is the effect of alcohol consumption in gout?
   - ➢ Alcohol metabolism leads to increased production of NADH, that enhances the conversion of monosodium urate to uric acid which is less soluble and can lead to precipitation of uric acid.
11. Give the causes of hyperuricemia in Von Gierke's disease.
   - ➢ In Von Gierke's disease G-6 phosphatase is deficient, so Glucose-6-phosphate cannot be converted to glucose. Accumulated G-6-P is channeled to HMP shunt where pentoses are formed (ribose). Increase in ribose sugars leads to increased formation of PRPP and purines. Increase in purine synthesis also increases degradation leading to increased uric acid production.
12. Name the enzymes of salvage pathway.
   - ➢ HGPRT (hypoxanthine guanine phosphoribosyltransferase).
   - ➢ APRT (adenine phosphoribosyl transferase).
13. What is Lesch Nyhan syndrome?
   - ➢ It is the disorder of salvage pathway caused due to deficiency of HGPRT. Features include hyperuricemia, renal stones and more importantly self mutilation (patient bites his own body parts like fingers).
14. What are causes of hypouricemia?
   - Xanthine oxidase deficiency.
   - Liver disease.
   - Renal tubular defects.
15. Why cancer is often associated with hyperuricemia?
   - ➢ In cancer there is increased cell growth as well as cell turnover resulting in increased catabolism of purine nucleotides. Purine catabolism produces uric acid.

Protein

1. What are normal values of plasma protein and A:G ratio?
   a. Total protein : 6-8 gm/dl
   b. Albumin (A): 3.5-5.5 gm/dl
   c. Globulin (G): 2-3.5 gm/dl

d. Fibrinogen : 0.2-0.4 gm/dl
e. A:G ratio:1.5-2.5

2. What are causes of hypoproteinemia or hypoalbuminemia?
   a. Kwashiorkor (decreased intake)
   b. Liver cirrhosis (decreased synthesis)
   c. Nephrotic syndrome (increased urinary loss).
   d. Fever, hyperthyroidism (increased catabolism).
   e. Hemodilution (due to water overload, prolonged steroid intake, diabetes insipidus).
3. What is the effect of hemodilution and hemoconcentration on A:G ratio?
   - A:G ratio does not change during hemodilation and hemoconcentration.
   - Hemodilution indicate false decrease in plasma protein due to increased in plasma volume whereas hemoconcentration indicate false increase in plasma protein due to decrease in plasma volume. False decrease or increase in protein can be thus confirmed by A:G ratio.
4. What is hypoalbuminemia?
   - Decrease in serum albumin to less 2 gm/dl.
5. What are causes of hyperproteinemia?
   a. Chronic infections
   b. Multiple myeloma
   c. Collagen diseases
   d. Hyperthyroidism etc.
6. What is multiple myeloma?
   - Malignant disease of plasma cells.
7. What is Bence Jones protein?
   - It is γ-globulin light chain excreted in urine of patient with multiple myeloma.
8. What is the feature of Bence Jones protein?
   - When heated, it is precipitated at 50-60°C. On further heating to the 100^{0}C precipitate dissolves and reappears on cooling.
9. What is the pattern of serum electrophoresis in multiple myeloma?

Start (−)

γ- globulin
M-band
β_2-globulin — mobility
β_1-globulin
α_2-globulin
α_1-globulin
Albumin (+)

Extra M-band is seen between γ-globulin and β-globulin band.

10. What is the significance of A:G ratio?
    - Normal A:G ratio:1.5-2.5
    - Decrease of A:G ratio: Renal or hepatic disease.
11. What is the clinical manifestation of hypoalbuminemia?
    - ➢ It causes edema. Albumin is the most abundant plasma protein so it contributes more towards the maintenance of osmotic pressure. When albumin level decreases, there is accumulation of fluid in the interstitial spaces leading to edema.
12. Define an albuminemia.
    - ➢ Absence of albumin synthesis.
13. What is microalbuminuria and proteinuria?
    - ➢ Normal protein excretion in urine=30-150 mg/day (i.e. 30 mg of albumin or 150 mg of protein per day).
    - ➢ Microalbuminuria: Excretion of small amount of albumin in urine (30-300 mg of albumin/day)
    - ➢ Proteinuria: Excretion of large amount of proteins in urine (more than 300 mgs to certain gms/day).
    - ➢ In nephrotic syndrome >3 gm of protein per day is excreted in urine.
14. What are other methods of protein estimation?
    - ➢ Kjeldahl's method (estimate nitrogen content of protein).
    - ➢ Lowry's method: (100 times sensitive than biuret method). It is used to estimate small amount of protein in fluids like CSF.
    - ➢ Turbidimetric method: Very sensitive method and is commonly used in CSF protein estimation.

Bilirubin

1. What is bilirubin? Give the types and normal range.
    - ➢ Bilirubin is catabolic end product of heme. It is of two types:
        a. Conjugated bilirubin
        b. Unconjugated bilirubin
    - ➢ Normal range:
        - Total bilirubin (conjugated+unconjugated) =0.2-1 mg/dl
        - Conjugated (direct bilirubin)=0.1-0.4mg/dl
        - Unconjugated (indirect bilirubin)=0.1-0.6 mg/dl
2. What are bile pigments?
    - ➢ Bilirubin → Yellow color
    - ➢ Biliverdin → Green color
3. Where does conjugation occur?

- It occurs in liver. Unconjugated bilirubin is water insoluble. It is transported to liver in bound form to albumin. In liver it conjugates with glucuronic acid to form conjugated bilirubin which is water soluble.

4. What is jaundice?
   - It is yellowish coloration of conjunctiva mucosa and skin due to increase in bilirubin level.
   - If bilirubin > 2 mg/dl, there is yellowish discoloration (visible symptoms) and this is known as clinical jaundice.
   - At the level of serum bilirubin 1-2 mg/dl, there are no visible symptoms like yellowish discoloration though the bilirubin level is high. This is known as latent jaundice.
5. What are the types of jaundice?

   i. **Acquired Jaundice**
   - *Physiological neonatal jaundice:* Occurs due to immature hepatic enzyme (glucuronyl transferase) system of newborns.
   - *Pathological neonatal jaundice:* Occurs due to Rh incompatibility at the time of birth.
   - Hemolytic/pre hepatic jaundice: Occurs due to increased hemolysis (hemolytic anemia, mismatch blood transfusion etc).
   - *Hepatic jaundice:* Occurs due to liver damage.
   - *Obstructive jaundice/posthepatic:* Occurs due to obstruction to flow of bile. Obstruction may be intrahepatic or extrahepatic. Causes include gallbladder stones, cancer of head of pancreas etc.

   ii. **Congenital Hyperbilirubinemias**
   - *Criggler Najjar syndrome (CNS)*: It is of two types:
     - CNS type I and CNS type II
       a. CNSI is due to severe deficiency of UDP glucuronyl transferase (bilirubin > 20 mg/dl).
       b. CNSII is due to mild deficiency of UDP glucuronyl transferase (bilirubin < 20 mg/dl).
       c. Unconjugated bilirubin level increases.
   - *Gilbert syndrome:* There is defect in uptake of bilirubin by liver. Unconjugated bilirubin is increased upto 3-5 mg/dl.
   - *Dubin-Johnson syndrome:* It is due to defect in ATP dependent anion transport protein. Liver cannot export the conjugated bilirubin, so that there is black staining of liver due to bilirubin deposition.
   - *Rotor syndrome:* It is caused due to defect in hepatic storage of bilirubin but there is no staining of liver.
6. What is known as black liver disease?
   - Dubin-Johnson syndrome.
7. What are other names for extrahepatic obstructive jaundice and hemolytic jaundice?
   - *Obstructive jaundice (extrahepatic):* Cholestatic jaundice.
   - *Hemolytic jaundice:* Acholuric jaundice.

8. What can be done for treatment of neonatal jaundice?
   - Phototherapy can be given. Light converts bilirubin to non-toxic lumirubin that can be easily excreted. This is known as photoisomerization.
9. What is kernicterus?
   - When bilirubin level is more then 18-20 mg/dl, it can cross the blood brain barrier and stain the brain yellow. This condition is known as kernicterus.
10. What do you mean by α, β, γ and δ bilirubin?
    - α bilirubin→Unconjugated bilirubin.
    - β bilirubin→Bilirubin monoglucuronide (20%).
    - γ bilirubin→Bilirubin diglucuronide (80%).
    - δ bilirubin→Bilirubin irreversibly bound to protein.
11. What is urobilinogen?
    - Conjugated bilirubin are deconjugated and converted to urobilinogen by intestinal bacteria.
12. What causes color in urine and feces?
    - Urobilinogen and stercobilinogen are colorless. They are produced from bilirubin by the action of intestinal bacteria. In presence of atmospheric oxygen, they are converted to urobilin and stercobilin which are colored. Urobilin gives characteristic pale yellow color to urine and stercobilin gives brown color to feces.
13. What is normal excretion of urobilin and stercobilin?
    - Urobilin → < 4 mg/day
    - Stercobilin → 300 mg/day
14. Is unconjugated bilirubin excreted in urine?
    - No, because it is water insoluble. Only water soluble conjugated bilirubin appears in urine (in case of obstructive jaundice). Normally it is not excreted. In obstructive jaundice, due to obstruction in flow of bile, it is regurgitated to systemic circulation. Conjugated bilirubin and bile salts being water soluble are excreted in urine.
15. Why the bilirubin is called as direct or indirect?
    - Conjugated bilirubin directly reacts with diazotized sulfanilic acid to produce color. So, it is called direct bilirubin.
    - Unconjugated bilirubin being water insoluble should be first converted to soluble form by adding methanol. Only then it will react with the diazotized sulfanilic acid. So, it is known as indirect bilirubin.
16. What is the method to estimate bilirubin?
    - Melloy and Evelyn method: It follows Van den Berg reaction for estimation of direct and indirect bilirubin.

Cholesterol

1. What is cholesterol?

- It is a sterol/derived lipid with 27 Carbon atoms.

2. Give the normal range of cholesterol?
   - 150-200 ml/dl.
   - One-third is present in non-esterified form (30%).
   - 70% is present in esterified form. Cholesterol is esterified by cholesterol esterase.
3. What is the significance of blood cholesterol?
   - Increased blood cholesterol is one of the important risk factors for atherosclerosis and myocardial infarction.
4. What do you mean by total cholesterol?
   - Total cholesterol= HDL+ LDL+ VLDL.
5. What is the normal level of triglyceride?
   - 50-150 mg/dl.
6. What are normal level of HDL, LDL and VLDL?
   - HDL=35-65 mg/dl
   - LDL=60-130 mg/dl
   - VLDL=20-40 mg/dl
   - VLDL=TG/5 (TG=50-150 mg/dl)
   - And LDL =TC-(HDL+VLDL)
7. Why are poly unsaturated fatty acids (PUFA) important in diet? Give sources.
   - PUFA enhance LDL uptake and their catabolism, thus they reduce the risk of coronary artery disease.

   Source: Vegetable oils (except coconut oils).
8. Give examples of some compounds formed from cholesterol.
   - Vitamin D
   - Bile acids
   - Steroid hormone
9. What is the nucleus present in cholesterol?
   - Cyclopentanoperhydrophenanthrene ring.
10. Give the name of bile acids?
    - Primary bile acid: Cholic acid and chenodeoxy cholic acid.
    - Secondary bile acids: Lithocholic acid and deoxycholic acid.

    Secondary bile acids are formed from primary bile acids.
11. What is known as good and bad cholesterol?
    - HDL is good cholesterol as it transports cholesterol from peripheral tissue to liver for excretion.
    - LDL is bad cholesterol as it transports cholesterol from liver to peripheral tissue for deposition. It increases risk of coronary artery disease and myocardial infarction.

12. What are the methods for cholesterol assay?

   i. **Chemical method:**
      - Zak's method.
      - Salkowiski's method.
      - Leibermann-Burchard method.
      - Sackett's method.
      - Kim and Goldberg method.

   ii. **Enzymatic method:** Cholesterol oxidase-peroxidase method (very sensitive).

13. What type of sample is used for cholesterol or lipid estimation?
   - ➢ Overnight fasting sample is used as dietary cholesterol tends to increase the serum cholesterol value.
   - ➢ Hemoglobin interferes colorimetric assay of cholesterol, so hemolyzed sample should not be used.
   - ➢ Alcohol consumption, exercise etc also affects lipid level, so these must be avoided prior to sample collection.

14. What are the tests done under lipid profile?
   - Total cholesterol (TC)
   - Triglyceride (TG)
   - HDL (high density lipoprotein)
   - LDL (low density lipoprotein)
   - VLDL (very high density lipoprotein)

15. What is cardiac risk ratio?

$$\textit{Cardiac risk ratio (CRR)} = \frac{\textit{Total cholesterol}}{\textit{HDL cholesterol}}$$

Male	Female	Risk
<1 – 3.5	<1 – 3.3	Below average
3.5 – 5	3.4 – 4.4	Average
5.1 – 9.6	4.5 – 7.1	Above average (2X)
>9.6	>7.2	Higher (3X)

16. What is Friedwald's equation? Give limitations.

LDL cholesterol=Total cholesterol – (HDL+ VLDL)

$$VLDL = \frac{Triglyceride}{5}$$

The Friedwald's equation should not be used under the following circumstances:
- ➢ When chylomicrons are present.
- ➢ When plasma triglyceride concentration exceeds 400 mg/dl (4.52 mmol/L).
- ➢ In patients with dysbetalipoproteinemia (type III hyperlipoproteinemia).

Appendix VI

NORMAL RANGES OF ANALYTES IN BLOOD

Analyte	Units	SI units
Ammonia	10-60 μg/dl	7.1-42.8 μmol/L
Acid phosphatase (ACP) total	1-3.5 KAU/dl	2.5-12 IU/L
Alanine aminotransferase		5-40 IU/L
Albumin	3.5-5.5 g/dl	35-505 g/L
Alkaline phosphatase	4-17 KAU/dl	40-125 IU/L
Acid phosphatase	1-3.5 KAU/dl	2.5-12 IU/L
Amylase	80-180 somogyi units/dl	50-120 IU/L
Aspartate aminotransferase	-	5-40 IU/L
Bicarbonate	21-28 mEq/L	21-28 mmol/L
Bilirubin (total)	0.2-1 mg/dl	4-17 μmol/L
Bilirubin (Direct)	0.1-0.4 mg/dl	<7 μmol/L
Bilirubin (Indirect)	0.1-0.6 mg/dl	<12 μmol/L
Calcium	9-11 mg/dl	2.1-2.5 mmol/L
Chloride	90-106 mEq/L	90-106 mmol/L
Cholesterol (total)	150-200 mg/dl	4-6 mmol/L
HDL-C	35-65 mg/dl	0.75-1.95 mmol/L
LDL-C	60-120 mg/dl	
VLDL-C	20-40 mg/dl	-
Copper	70-150 mg/dl	11-24μmol/L
C-reactive protein (CRP)	0.5-1 g/dl	-
Creatine	0.2-0.4 mg/dl	15-30μmol/L
Creatine kinase	10-50 IU/L	10-50 IU/L
Creatinine	0.7-1.5 mg/dl (male) 0.6-1.1 mg/dl (female)	60-125 μmol/L 35-106 μmol/L
Creatinine clearance	95-140 ml/min (male) 85-130 ml/min (female) Average:125 ml/min	-
Fibrinogen	200-400 mg/dl	12-42 μmol/L
Globulins	2.5-3.5 g/dl	25-35 g/L
Glucose (fasting)	60-100 mg/dl	4-6.1 mmol/L
Glucose (post prandial)	100-140 mg/dl	6.1-8.5 mmol/L
Glucose (random)	60-140 mg/dl	4-8.5 mmol/L
Hemoglobin	14-16 gm/dl (male) 13-15 g/dl (female)	2.17-2.4 mmol/L
Iodine	5-10 μg/dl	-
Iron	50-160 mg/dl	9-28.6 mmol/L
Total Iron binding capacity	250-400 mg/dl	45-80 mmol/L
Lactate dehydrogenase	140-330 IU/L	140-330 IU/L
pCO_2	34-46 mm Hg	-
pO_2	90-110 mm Hg	-
Phosphate	2.5-4.5 mg/dl	1-1.5 mmol/L
Potassium	3.5-5 mEq/L	3.5-5 mmol/L

Prostate specific antigen	100-500 ng/dl	1-5 μg/L
Protein (total)	6-8 g/dl	60-80 g/L
Sodium	136-145 mEq/L	136-145 mmol/L
Total T_3(Tri-iodothyronine)	75-190 ng/dl	1.16-3 nmol/L
Total T_4 (Thyroxine)	5-12 μg/dL	65-150 nmol/L
Free T3 (fT3)	2.3-6.6 pg/ml	3.5-10 pmol/L
Free T4 (fT4)	0.8-2.3 ng/dl	10.2-29.4 pmol/L
T_g (Thyroglobulin)	3-5 μg/dl	3-50 μg/L
TSH	0.5-5 μU/ml	0.5-5 mU/L
Transferrin	200-300 mg/dl	23-35 μmol/L
Triglycerides	50-150 mg/dl	0.5-1.3 mmol/L
Urea	20-40 mg/dl	2.4-4.8 mmol/L
Urea clearance	Maximal=75 ml/min Standard=54 ml/min	-
Urea nitrogen	8-20 mg/dl	3-9 mmol/L
Uric acid	3-7 mg/dl	0.18-0.35 mmol/L
Vitamin A	15-50 μg/dl	12-42 μmol/L
Vitamin C	0.4-1.5 mg/dl	12-42 μmol/L
Vitamin D_3	1.5-6 μg/dl	50-160 pmol/L
Vitamin E	0.5-1.8 mg/dl	12-42 μmol/L

Normal Ranges of Analytes in Urine

Analyte in urine	Normal value
Total protein	<150 mg/day
Albumin	< 30 mg/day
Creatinine	1.5-2 gm/day (male) 0.8-1.5 gm/day (female)
Urea	2 gm/dl or 30 gm/day
Amylase	1500-1800 Somogyi units/day
Chloride	10-15 gm/day as NaCl

Normal Ranges of Analytes in CSF (Cerebrospinal Fluid)

Analyte in CSF	Normal value
Protein	15-45 mg/dl
Sugar	
Chloride	120-130 Eq/L

Normal Ranges of Analytes in Gastric Juice

Basal acid output (BAO)	1-2.5 mmol/hr
Maximal acid output (MAO)	20-40 mmol/hr